Palgrave Studies in Economic History

Series Editor
Kent Deng, London School of Economics, London, UK

Palgrave Studies in Economic History is designed to illuminate and enrich our understanding of economies and economic phenomena of the past. The series covers a vast range of topics including financial history, labour history, development economics, commercialisation, urbanisation, industrialisation, modernisation, globalisation, and changes in world economic orders.

Martha Emilie Ehrich

The Dutch Paper Industry from 1580 to the Present

The Survival of a Long-Established Industry

palgrave
macmillan

Martha Emilie Ehrich
Film University Babelsberg
KONRAD WOLF
Potsdam, Germany

ISSN 2662-6497 ISSN 2662-6500 (electronic)
Palgrave Studies in Economic History
ISBN 978-3-031-54323-4 ISBN 978-3-031-54324-1 (eBook)
https://doi.org/10.1007/978-3-031-54324-1

This publication was supported by funds from the Publication Fund for Open Access Monographs of the Federal State of Brandenburg, Germany. It was also supported by the Publication Fund of the Film University Babelsberg KONRAD WOLF.

This Palgrave Macmillan imprint is published by the registered company Springer Nature Switzerland AG
The registered company address is: Gewerbestrasse 11, 6330 Cham, Switzerland

Paper in this product is recyclable.

Dedicated to Angela

Praise for *The Dutch Paper Industry from 1580 to the Present*

"*The Dutch Paper Industry from 1580 to the Present* is not your typical history book. By taking a historical materialist approach to expose the complicity between capitalism and networks, Martha E. Ehrich makes an invaluable contribution to current economic history scholarship. Utilizing the century-old craft of paper making in The Netherlands as an empirical case, the book guides readers through every stage of the development of this industry, providing a captivating explanation of how and why capitalist class fractions intersect in networks to navigate the (in-)stability of industrial capital. This is essential reading for everyone who is interested in critical analyses of the global paper industry and the history of capitalist development more generally."

—Professor Steffen Boehm, *Professor in Organisation and Sustainability at the University of Exeter Business School, UK*

Contents

List of Figures

List of Tables

CHAPTER 1

Networked Industry Survival

Abstract This chapter introduces the reader to the deeply entrenched relation of capitalist networks and the Dutch paper industry. The production of paper from rags—old and torn cloth—already existed in pre-industrial seventeenth-century Netherlands and continued well into the nineteenth century. Initially, the chapter conducts an international comparison of this well-established industry before delving into a more detailed exploration of the four dimensions of networks in capitalism: technology, state-industry relations, competition and cooperation, and labor-capital relations. The chapter concludes by outlining the methodology used to historicize the industry across four phases of capitalism: the rise of Dutch capitalism (1580–1815), Dutch monarchic liberalism (1815–1914), Fordism (1914–1980), and post-Fordism (1980 until now).

Keywords Dutch paper industry · International comparison · Networks in capitalism · Historicization · Rags

RAGS make paper,
  PAPER makes money,
  MONEY makes banks,
  BANKS make loans,
  LOANS make beggars,
  BEGGARS make RAGS.

M. E. Ehrich, *The Dutch Paper Industry from 1580 to the Present*, Palgrave Studies in Economic History,
https://doi.org/10.1007/978-3-031-54324-1_1

Author unknown, around the eighteenth century.[1]

This eighteenth-century anonymously authored poem captures the deeply entrenched relation of capitalism and paper technologies. Paper not only shaped capitalist activities through, for example, the rise of accounting practices in organizations and the use of maps for colonial projects but also served as metaphor and materiality of capitalist exchange and wealth extraction through, for example, the introduction of paper money.[2] The production of paper from rags—old and torn cloth—already existed in the pre-industrial seventeenth-century Netherlands and continued well into the nineteenth century. Initially introduced from the Islamic world in the twelfth century, papermaking gradually gained momentum and transformed into a vital industry in Europe.[3] Prior to paper, parchment and vellum, made from animal skins, were the primary writing materials. However, paper proved to be a more affordable and versatile alternative.[4]

Paper can be described as a watery accumulation of various plant fibers and other substances, arranged in the shape of a sheet or web. It is formed from a fibrous material that arises from the extraction of plant fibers through mechanical methods, chemical methods, or a combination of both. During the paper production process, these liberated fibers are reassembled either using a box-like mold featuring fabric bottoms and wire meshes or later through a paper machine.[5] Rags proved to be a superb substance for paper production because a solitary cotton fiber held remarkable robustness, enabling it to endure substantial weight despite its relative thinness.[6] The origins of Dutch paper production can be traced back to the utilization of rags.

Research on paper production and trade is currently experiencing a vibrant surge within the academic world, as researchers endeavor to

[1] Quoted in Hunter (1978). According to Senchyne (2017, p. 545) "All known citations of this poem originate with Hunter, who does not cite his source except to say that the author is anonymous and it is from the late-eighteenth-century United States".

[2] Senchyne (2017, p. 546).

[3] Basbanes (2013, p. 11) and Bellingradt (2019, p. 68).

[4] Basbanes (2013, p. 169).

[5] Särkkä et al. (2018, p. 2).

[6] Ibid.

address an important historiographical gap. With advancements in interdisciplinary studies and access to diverse historical sources, scholars are delving into the intricacies of papermaking techniques, the development of paper mills, and the dissemination of paper across different regions and cultures.[7] This increased attention stems from the recognition that the history of paper is integral to understanding the evolution of knowledge, communication, and cultural exchange throughout human civilization. To shed light on the profound impact of paper on various aspects of society, recent studies on paper production and dissemination contribute to a more nuanced and comprehensive understanding of the historical significance and transformative power of this ubiquitous material.[8] This book contributes to this renewed interest in paper production by delving into the historical evolution of the Dutch paper industry[9] over a span of six centuries since its inception.

1.1 A Long-Established Industry in International Comparison

The Dutch paper industry evolved under state monopoly capitalism in the Seven United Provinces from 1588 onwards. Commonly extenuated as the 'Golden Age', this phase describes seventeenth-century Dutch wealth creation through colonial exploitation.[10] During this time, the Dutch paper industry became internationally famous for its superior product quality. This superior quality was achieved through distinctive production methods and the disposability of raw materials to Dutch paper makers, first and foremost rags, and, crucially, water and wind. Due to the Netherlands' flourishing textile production and shipping industry at that time, rags and cloth were readily and affordably accessible to paper makers in the region.

[7] See for example: Bellingradt (2017, 2019), Bloom (2001), Chang and Andreoni (2021), and Särkkä (2021).

[8] Bellingradt and Reynolds (2021).

[9] Initially focused solely on paper production, the Dutch paper industry diversified to include board, a thicker paper, in the late seventeenth century, and later expanded to manufacture cardboard, an even thicker and more rigid type, in the nineteenth century. As a result, certain scholars opt to refer to the industry as the Dutch paper and board industry.

[10] Arrighi (2010, p. 153).

The introduction of the *Hollander* beater marked a significant advancement in Dutch papermaking, enhancing the reputation of Dutch paper production even more. This machine, featuring multiple bronze blades rotating on a wooden or metal foundation, was utilized to meticulously beat rags and extract their fibers. Even though the resulting quality of paper was much lower than that produced by a wood-pounder (*hamerbak*) in terms of strength and contaminants, Dutch paper makers had gained worldwide fame with their largely increased production output by the end of the seventeenth century.

Around this time, paper mills were set up all across Europe. Next to the Netherlands, Germany also established itself as one of the leading centers of paper production in Europe due to its favorable geographic conditions, abundant water resources, and access to raw materials.[11] Driven by a combination of trade, technological innovations, and a growing demand for paper, water-powered mills, stamping technologies, and refining equipment furthered the quality and efficiency of paper production in Germany. Especially printing and publishing industries started flourishing here, leading to the increasing publication of books and manuscripts.[12]

The Dutch paper industry also played a crucial role in the cultural development at that time. With its ability to produce high-quality paper, the industry met the demand for books, maps, and other printed materials across Europe and beyond. In fact, the Dutch paper industry's reputation for producing high-quality paper fostered a vibrant publishing and printing sector in the Dutch Republic, too. Dutch printers and publishers capitalized on the availability of paper, contributing to the dissemination of knowledge, advancements in printing technology, and cultural production.

The extensive utilization of paper for printing scientific texts and treatises, along with the ensuing dissemination of knowledge, played a pivotal role in fostering the Scientific Revolution during the sixteenth and seventeenth centuries in Europe. Intellectuals and academics made notable advancements across diverse fields during this period. Paper became an essential medium for preserving and transmitting scientific discoveries, enabling the accumulation of knowledge and the development of scientific

[11] Turunen (2012, pp. 82–83).

[12] Ibid.

disciplines.[13] Additionally, paper's affordability and versatility also played a significant role in the cultural realm, as it facilitated the production of literature, poetry, and artistic works, contributing to the flourishing of cultural heritage.[14]

In contrast to the Netherlands and Germany, the development of papermaking in Scandinavia was characterized by a gradual adoption and adaptation of the craft between the fourteenth and seventeenth centuries.[15] Relying on imported paper and raw materials primarily from Germany and the Netherlands, Scandinavia strongly depended on its trading partners. England, on the other hand, relied on its own paper mills and raw materials from the seventeenth century onwards.[16] Similar to the development in the Netherlands, England's paper industry was fueled by advancements in technology, by the acceleration of colonialism, and by the growing demand for paper. Water-powered mills became the primary method of production here too, utilizing the power of rivers and streams to drive the machinery.[17]

Soon the worldwide paper demand resulted in a scarcity of rags and a subsequent increase in their prices throughout Europe. From the mid-eighteenth century onwards, European papermakers started searching for a more abundant, cost-effective raw material. From 1840 onwards the main ingredient of paper—rags—was replaced by wood fiber. Making paper from wood fiber necessitates pulping. Pulping is a process, which transforms fibrous raw materials into a collection of liberated fibers, achieved by dissolving the binding components (primarily lignin) that hold the cellulosic fibers together.[18] Pulping can include mechanical beating and/or chemical methods and demands a lot of energy. As a result, the emergence of wood fiber as a primary material for paper production led to a significant rise in the importance of forestry, substantial energy sources, and the supply of chemicals for paper manufacturers.

Around this time, Scandinavia transitioned from its reliance on imported paper and raw materials from Germany and the Netherlands

[13] Bellingradt (2019, p. 11ff.).

[14] Ibid., pp. 77–79, 85–86.

[15] Järvinen et al. (2012, pp. 20–21).

[16] Särkkä (2021)

[17] Särkkä (2012, p. 169).

[18] Särkkä et al. (2018, pp. 3–4).

to establishing its own raw material supply and thriving paper mills.[19] The abundant water resources and vast forests in Scandinavia provided favorable conditions for paper production.[20] Not only Scandinavian paper makers but also those in Central Europe and North America had access to large quantities of forest, including highly sought-after coniferous trees like spruce and fir trees, which were optimal to deliver highest quality fibers for paper production. Also Britain adapted quickly, as the expansion of its colonial empire entailed the extraction of vast wood areas, particularly in North America, ensuring a steady supply of raw materials for British paper production.[21]

In the nineteenth century, there was a significant increase in the production and circulation of newspapers in Europe compared to the previous centuries. The rise of industrialization, urbanization, and advancements in printing technology led to a proliferation of newspapers catering to various interests and demographics. Consequently, during the nineteenth century in Europe, the production of newsprint experienced a significant increase in proportion to previously popular paper types such as uncoated paper stock and stationery. The Netherlands lacked the substantial quantities of pulp required for paper production, including newsprint. The nation once had a thriving forest dominated by hardwoods, especially oak trees. However, this valuable resource dwindled over time due to widespread deforestation for agricultural purposes, firewood, and population expansion, reaching a critical point by the conclusion of the Middle Ages. Due to the enduring shortage of native coniferous trees necessary for pulp production, Dutch papermakers tried to either hold onto traditional papermaking materials or explore alternative raw materials like straw.

Around the same time as wood fiber was introduced, industrialization took off in several European countries, and innovations such as the Fourdrinier machine enabled the continuous production of paper in large rolls, propelling production capacities in Britain, Belgium, Scandinavia, Germany, and France even further.[22] Between 1900 and 2005, paper machines' top speed increased by a factor of ten from 200 to

[19] J. Järvinen et al. (2012, pp. 20–21).

[20] Ibid.

[21] Särkkä (2012, p. 174).

[22] Ibid., p. 170.

2,000 m/min, and machines are nowadays more than three times wider, with a width of up to 11 meters.[23] Since 1900, worldwide paper production increased 40-fold to 409 million tons in 2016.[24] This period also witnessed a diversification of paper types, including newsprint, packaging materials, and specialty papers, establishing both Scandinavia and Britain as significant players in the global paper market.

The swift growth of the paper and pulp industry in Scandinavia, Central Europe, and North America far outpaced the progress of the Dutch paper sector. Dutch paper makers tried to keep up by finally transitioning from traditional handmade paper techniques to large-scale, industrialized production by the end of the nineteenth century. Around that time, the Dutch state made its first attempts at forest cultivation, particularly monoculture pine plantations.[25] However, in addition to significant technical and economic competition, obstacles resulting from the geographical location and conditions of the Netherlands proved insurmountable for Dutch paper producers. Scandinavian producers, for instance, capitalized on their inherent strengths, which included ample pulp resources, cost-effective hydropower, extensive waterway networks, and favorable geographical proximity. They further solidified their position by employing pricing strategies and establishing marketing cooperatives like the Central Association of the Finnish Forest Industry. Through such initiatives, Finnish forest companies established a firm grip on, among others, the Dutch market.[26]

All these factors had a profound impact on the profitability and international standing of the Dutch paper industry. With increasing foreign competition, the dominance of Dutch paper in Europe started to decline rather quickly. Due to the prolonged transition from traditional handmade paper techniques to large-scale industrialized production in the Netherlands, coupled with its geographical disadvantage, the Dutch paper industry found itself in a fixed position in the global modern pulp and papermaking landscape. Essentially, the Netherlands was never optimally situated to emerge as a substantial or enduring contender in the modern pulp and paper sector. While it could temporarily address challenges posed

[23] Särkkä et al. (2018, pp. 3–4).

[24] Diesen (2007, p. 99).

[25] Oosthoek (2018, p. 68).

[26] Jensen-Eriksen and Ojala (2015, p. 527).

by its geographical location and environmental conditions, it couldn't avoid the inevitable outcome. The scarcity of local raw materials, especially coniferous fibers, put Dutch papermakers at a distinct disadvantage. Despite their efforts to manage by importing pulpwood, pulp, and paper supplies, this disadvantage persisted for an extended period.

Left behind with non-competitive, niche-focused paper products, the history of the Dutch paper industry is marked by a decline in production sites and machines. Nonetheless, industrialization also helped increase production output since. In 1740, the Dutch paper industry was comprised of roughly 150 paper mills in the region of Veluwe, a forest and heather area in central Netherlands, and 40 paper mills in the region of Zaanstreek, an industrial area in North Holland, producing a total of 3,750 tons of paper annually.[27] Almost three centuries later in 2021, 23 paper mills produced a total of 2,942,000 tons of paper annually in the Netherlands.[28]

While the overall decline in a number of paper mills has been a longer-standing reality for the Dutch paper industry, the industry managed to survive. The industry's resilience becomes particularly striking when considering its international context: No longer the leading European paper-producing country as it was during the sixteenth and early seventeenth century, it never rose back to its old fame. Presently, the Dutch paper industry is recognized as a small sector in relation to both its production volume and the dimensions of its paper mills when juxtaposed with global rivals. This compactness enables the industry to foster innovation and engage in exploratory ventures involving novel paper manufacturing methods and products. Unlike its international counterparts that prioritize incessant growth in paper output, the Dutch paper sector stands out as a specialized, small-scale, and innovative sector focused on high-quality paper and board production. Today, the historical legacy of the Dutch paper industry as a hub of innovation and trade remains an important part of the country's industrial heritage and cultural identity.

To gain a deeper understanding of how this long-established industry managed to survive six centuries of capitalist development, this book foregrounds the crucial role of cooperative networks in alleviating some of the

[27] Bouwens (2004, pp. 23, 29).

[28] VNP (2021, pp. 48–49).

pressures arising from the contextual challenges faced by the Dutch paper industry. Networks were and continue to be essential to the industry's development throughout capitalism. Central to maintaining the profitability of the Dutch paper industry over time, networks played a crucial role in facilitating the negotiation of industrial interests within the context of four key dimensions: State-industry relations, capital-labor relations, competition and cooperation, and technology. Although these dimensions are frequently analyzed independently, it is the combined interaction of networks spanning these four dimensions that markedly bolstered the industry's enduring strength over the course of its history.

These networks played a pivotal role in enabling the industry to adeptly embrace technological progress, effectively manage and alleviate labor disputes, navigate the intricacies of competitive forces by fostering intra-industry cooperation and aligning with governmental bodies, as well as advocate for the interests of key industrial stakeholders. It is precisely along, within, and between these four dimensions that different class fractions were able to negotiate and install their interests as general interests on both national and international levels as well as within the Dutch paper industry, culminating in those networked relations, which satiated industrial capital and safeguarded the industry's survival, albeit in ever-more marginal form.

1.2 Networks in Capitalism

Painting a picture of a long-established industry and its survival throughout different phases of capitalism, this book captures the paramount importance of various networks between state, industry, and labor agents to the industry's viability. Previous historical research has shown that sets of colonial, trade, merchant, and family networks, tightly interwoven through a dense web of capital, were pivotal to paper production and trade in early modern Europe.

First, colonial networks formed a crucial aspect of paper production and trade as colonies were exploited and extracted for raw materials, like cotton and rags, as well as enslaved labor. These networks facilitated the extraction of raw materials and their transportation back to Europe, ensuring a steady supply for the growing demand of the paper industry.[29] Additionally, trade networks, such as the Dutch East India Company (VOC), were key players in the colonial trade of goods, expanding the reach and influence of European paper globally.[30] These networks distributed paper across Europe through marine and land transportation, connecting colonial production hubs, harbors, paper mills, printing houses, and customers. Trade networks also facilitated the import of raw materials, like dyes and chemicals, necessary for paper production in the Netherlands, thereby supporting the industry's growth and diversification.

Also, merchant networks were instrumental in driving paper production and trade. Merchants formed connections with paper mills, printers, and customers, creating a network that facilitated the distribution and sale of paper.[31] They acted as intermediaries, negotiating contracts, arranging shipments, and establishing trade routes.[32] Merchant networks also played a vital role in financing paper production, providing capital for investments in equipment, expansion of production capacity, and marketing efforts. The expertise and connections of these merchants contributed to the expansion of paper production and the development of a competitive European trade market.

Lastly, family networks played an important role in the paper trade.[33] In smaller-scale paper production, family-based paper mills and workshops were common in the Dutch region called Veluwe, where knowledge and skills were passed down through generations. These family networks allowed for the transmission of specialized techniques, ensuring the production of high-quality paper. Family connections provided a source of labor, with family members working together in different aspects of the paper production process. In larger-scale production, family networks encompassed the transferal of generational wealth, such as ownership of paper mills, particularly in the region of Zaanstreek, and shares in other lucrative financial investment and colonial outlets. The cohesion and

[29] Antunes (2018).

[30] Birkenholz (2021).

[31] For further discussions of paper trade and merchant networks in England see: Da Rold (2020, p. 87ff.).

[32] Jonker and Sluyterman (2000).

[33] See for further information on the interplay of family and trade networks: van den Tol (2020, p. 15ff.) and Fusaro (2012).

continuity of both these family networks contributed to the stability and growth of paper production and trade in the Netherlands. Both merchant and family networks have been closely intertwined in the early modern Netherlands, as family patriarchs aided state-building processes through their merchant capital.[34]

To enrich the revitalized debates on the role of networks for paper production, this book foregrounds four key dimensions of networks throughout different phases of capitalism, namely state-industry relations, capital-labor relations, cooperation and competition, and technology. To do so, networks are understood as forms and fora of organization, which are historically contingent social phenomena and embedded in capitalist relations of (re-)production. Capitalism rests on both the social organization of production *and* reproduction of the means for human subsistence. While production refers to the exploitation of commodified labor by those owning the means of production for the accumulation of surplus value, reproduction refers to "the work required to maintain people as social, emotional, and intellectual beings on a daily and intergenerational basis".[35] While classical Marxists solely focus on the antagonism between capital and labor as constitutive of capitalist social relations, queer-feminist, anti-racist Marxists break with the idea that class is the only constitutive dimension of social relations of (re-)production in capitalism, instead pointing out how gender, sexuality, race, nation, and (settler) colonialism simultaneously determine these relations.[36]

Capitalist relations of (re-)production describe the economic and social order we live in, namely "the accumulation and competition of capitals".[37] In turn, these social relations of (re-)production constitute social relations of power within capitalist societies. As each capitalist phase is demarcated by a changing institutional setup of the state, which always reproduces capitalist relations of (re-)production, the state in capitalism is always a capitalist state.[38] Understanding industry development more generally and the survival of the Dutch paper industry more specifically, thus, becomes a matter of scrutinizing the changing relations between

[34] Adams (1994, p. 516); e.g. Fusaro (2012).

[35] Glenn (2010, p. 208).

[36] Nguyen (2023) and Ferguson (2008).

[37] Banaji (2010, p. 14ff.).

[38] Poulantzas (2000, p. 126).

industry and state. Since regulatory regimes in capitalism imply specific "constellation[s] of legal, administrative and coercive state apparatuses to both legitimize and shield themselves from political and social contestation", a historicization of the Dutch paper industry has to, therefore, entail an analysis of the particular spatial–temporal relations between state and capitalist class fractions.[39] In fact, the very processes of negotiating hegemonic ambitions between the political, administrative, and capitalist class fractions are the cohering and decaying forces of the networks that sustained the Dutch paper industry throughout different phases of capitalism.[40] Thus, the industry's networked survival needs to be traced along the lines of state-industry relations.

Capitalist relations of (re-)production are always already classed, gendered, and racialized relations. The state is "a historically determined form of the organization of domination".[41] This said, the state mediates the conflicts between different capital fractions, while labor and social struggle are subjugated within these.[42] As a matter of fact, the state's institutional setup conceals who the dominating class(es) and who the dominated classes are, due to framing itself as being representative of 'the people'. Political and material hierarchies in terms of gender, race, and class are hidden behind the concept of a democratic state, which in tendency universalizes dominant capitalist interests as the interests of all. Times of increased socio-political conflict occasionally expose this façade: In critical times, state institutions—as agents in their own right—no longer only rely on active or passive consent, but actively exert coercion.[43] Overall, the very processes of convulsion between state and labor (including classed, gendered, and racialized labor forces) are additional cohering and decaying forces of the networks that sustained the Dutch paper industry throughout different phases of capitalism. Thus, the industry's networked survival needs to also be traced along the lines of capital-labor relations.

[39] Bruff and Tansel (2018, p. 7).

[40] For a comparable analysis of business-government relations in the British paper-pulp industry between 1950 and 1980 please see: MacKenzie (2018).

[41] Wigger and Buch-Hansen (2013, p. 615).

[42] Poulantzas (2000, p. 125).

[43] Harvey (2003, p. 36ff.).

Processes of competition and cooperation are innate to capitalist relations of (re-)production. As exchange-value is privileged over use-value in capitalism, "cooperation and mutual aid—the antithesis to competition—are marginalized as organizing principles".[44] In fact, as cooperation becomes merely an effort to decrease competition, they become two sides of the same coin. The fact that capitalist accumulation regimes are anchored in competition to (re-)produce surplus value and that cooperation exists in a tensional relation to competition, leads to growing disintegration.[45] As a result, industries take on new forms of institutionalized power-play between various capitalist class fractions throughout time. These class fractions compete with each other, while at the same time aligning their common interests to maintain the capitalist system. Consequently, also processes of competition and cooperation cohere and decay the very networks that sustained the Dutch paper industry throughout different phases of capitalism. Thus, the industry's networked survival needs to be traced along the lines of competition and cooperation.

Capitalist relations of (re-)production also manifest in technology. As the material conditions of production, technology more generally, and technological innovation more specifically are crucial to capitalism. Capitalist class fractions continuously compete for profit maximization, which in turn drives the introduction of ever-more powerful and productive technology.[46] Since "technology is produced amidst [these] conflicting social relations",[47] it needs to be understood as both a material artifact as well as a discursive practice, being shaped by and itself shaping politico-economic realities.[48] Technological innovation, thus, not only effects industry performance, production processes, working conditions, consumer demands, and consumption levels but is itself constitutive of the networks that sustained the Dutch paper industry throughout different phases of capitalism. Thus, the industry's networked survival needs to be traced along the lines of technology.

[44] Wigger and Buch-Hansen (2013, p. 608).

[45] Ibid.

[46] Davis et al. (1997, p. 4).

[47] Ibid.

[48] Fisher (2010, p. 231).

1.3 (Re-)writing History

Capitalist history, and thus the history of the Dutch paper industry, emerges through distinct, yet successive phases of changing capitalist relations of (re-)production.[49] More often referred to as modes of regulation, meaning the social conditions under which (re-)production is organized, these phases are the manner as well as the object of regulation.[50] As history succeeds through continuous discontinuity, each mode of regulation struggles to serve as a supportive base to the respective accumulation regime in times of severe crisis. Retrospectively, one can identify meaningful conjunctures for the gradual succession of a particular mode of regulation by another one.

The rise of capitalism in the Netherlands started with the transition from the feudal to the capitalist mode of production, which was closely interlinked with the rise of Dutch colonial power during the sixteenth century and the subsequent development of national, capital markets, as well as secondary financial markets.[51] This co-called second rise of capitalism is rooted in the Dutch cycle of capital accumulation, sketching out the 'uniqueness' of Dutch capitalism (1580–1815).[52] This phase is furthermore paradigmatic of the non-linear transition from feudalism to capitalism, as feudally organized paper production in the Veluwe coexisted alongside more capitalistic organized paper production in the Zaanstreek.[53]

Dutch monarchic liberalism (1815–1914) describes the second, distinct phase of capitalism in the Netherlands due to the reinstallment of a monarchic rule in 1813.[54] Consequently, the emergence of the gentlemanly capitalist class fraction in the first half of the nineteenth century and the rise of Dutch liberals in the second half of the nineteenth century have to be understood in direct relation to the Dutch monarchy under William I (reigned 1813–1815) and William II (reigned 1840–1849).[55]

[49] Jessop (1993, 2002) and Jessop and Sum (2006).

[50] Hirsch (1978, p. 58) and Jessop (1994, p. 276).

[51] Anievas and Nisancioglu (2015, p. 180ff.).

[52] Arrighi (2010) and Brandon (2011, 2015).

[53] Federici (2004, p. 62).

[54] Te Velde (2008).

[55] Davids (2006), Kuitenbrouwer and Schijf (1998), and Schrauwers (2010).

In regards to the Dutch paper industry, this phase was marked by two major technological shifts in paper production, the shift from handmade paper to mechanically produced paper and the change from cloth to cellulose fibers.

With the beginning of the nineteenth century, Fordism (1914–1980) demarcates the third phase of Dutch capitalism, in which mass production technologies and Taylorist practices of organizing working procedures became predominant.[56] In the context of the Dutch paper industry, this third phase is primarily marked by the transition from virgin wood fiber to waste paper fiber in the 1950s. As the state took efforts to stimulate nationwide household paper recycling, waste paper material became cheaply available in the Netherlands, relieving Dutch paper producers from costly imports of virgin wood pulp. During this time, the visible hand of the state also played a major role in stimulating mergers, takeovers, and the internationalization of Dutch industrial sectors.[57] Furthermore, the 'old boys network', comprised of supervisory directors of big corporations and banks, started controlling the majority of Dutch industries.[58] These network relations abetted the later dominance of the financial class fraction, which profited from the industrial declines and crises of the 1970s through private and public bank loans.

Consequently, the following phase of Post-Fordism (1980–now) is defined by the rise of finance-led accumulation patterns. In addition, the transnationalization of production and the subsequent emergence of global value chains, deindustrialization, deregulation and neoliberal re-regulation, and labor market flexibilization all mark this fourth phase.[59] Often romanticized as a time of state retrenchment and free markets, post-Fordism actually involves new forms of state-led market intervention, for example, in the form of direct financial aid.[60] Concerning the Dutch paper industry, the co-evolution of the industry's network organization (KCPK) and the Dutch Competition Authority (NMa) is a prime

[56] Koch (2004, pp. 3–6).

[57] Van Zanden (2005).

[58] Horn and Vliegenthart (2010, pp. 63, 67).

[59] Jessop (1994), Jessop and Sum (2006), Overbeek et al. (2007).

[60] Panitch and Konings (2009).

example of the development of post-Fordist cooperation between governmental authorities and companies of the Dutch paper industry in a climate of hyper-competition.

The historicization of the Dutch paper industry—as well as the chapters in this book—are structured according to these four phases of capitalism: The rise of Dutch capitalism (1580–1815), Dutch monarchic liberalism (1815–1914), Fordism (1914–1980), and post-Fordism (1980–2016). Since the four phases are fluid in their timely and spatial boundaries, the periodization sketches out the specific industrial context of Dutch papermaking and, hence, remains non-generalizable.[61] Nevertheless, it can serve as a viable reference point for future research dedicated to offer similar, critical accounts of capitalist industrial history.

References

Adams, J. (1994). The familial state: Elite family practices and state-making in the early modern Netherlands. *Theory and Society, 23*(4), 505–539.

Anievas, A., & Nisancioglu, K. (2015). *How the West came to rule*. Pluto Press.

Antunes, C. (2018). From binary narratives to diversified tales. *Tijdschrift voor Geschiedenis, 131*(3), 393–407.

Arrighi, G. (2010). *The long twentieth century: Money, power and the origins of our times*. Verso.

Banaji, J. (2010). *Theory and history: Essays on modes of production and exploitation*. Brill.

Basbanes, N. (2013). *On paper: The everything of its two-thousand-year history*. Alfred A. Knopf.

Bellingradt, D. (2019). *Vernetzte Papiermärkte*. Herbert von Halem.

Bellingradt, D. (2017). Paper networks and the book industry: The business activities of an eighteenth-century paper dealer in Amsterdam. In D. Bellingradt, P. Nelles, & J. Salman (Eds.), *Books in motion in early modern Europe: Beyond production, circulation and consumption* (pp. 67–85). Palgrave Macmillan.

Bellingradt, D., & Reynolds, A. (Eds.). (2021). *The paper trade in early modern Europe: Practices, materials, networks*. Brill.

Birkenholz, F. (2021). The paper purchases of the Dutch East India Company's Amsterdam chamber in the early eighteenth century. In D. Bellingradt & A. Reynolds (Eds.), *The paper trade in early modern Europe: Practices, materials, networks* (pp. 244–265). Brill.

[61] Duménil and Lévy (2001).

Bloom, J. (2001). *Paper before print*. Yale University Press.

Bouwens, B. (2004). *Op papier gesteld: de geschiedenis van de Nederlandse papier- en kartonindustrie in de twintigste eeuw*. Boom.

Brandon, P. (2015). *War, capital, and the Dutch state (1588–1795)*. Brill.

Brandon, P. (2011). Marxism and the 'Dutch miracle': The Dutch Republic and the transition-debate. *Historical Materialism, 19*(3), 106–146.

Bruff, I., & Tansel, C. B. (2018). Authoritarian neoliberalism: Trajectories of knowledge production and praxis. *Globalizations*, 1–12.

Chang, H.-J., & Andreoni, A. (2021). Bringing production back into development: An introduction. *The European Journal of Development Research, 33*(2), 165–178.

Da Rold, O. (2020). *Paper in medieval England: From pulp to fictions*. Cambridge University Press.

Davids, K. (2006). The transformation of an old industrial district: Firms, family, and mutuality in the Zaanstreek between 1840 and 1920. *Enterprise and Society, 7*(3), 550–580.

Davis, J., Hirschl, T., & Stack, M. (1997). Introduction. Integrated circuits, circuits of capital, and revolutionary change. In J. Davis, T. Hirschl, & M. Stack (Eds.), *Cutting edge: Technology, information capitalism and social revolution* (pp. 1–10). Verso.

Diesen, M. (2007). *Economics of the pulp and paper industry: Papermaking science and technology*. Finnish Paper Engineers' Association.

Duménil, G., & Lévy, D. (2001). Periodizing capitalism: Technology, institutions and relations of production. In R. Albritton, M. Itoh, R. Westra, & A. Zuegge (Eds.), *Phases of capitalist development* (pp. 141–162). Palgrave Macmillan.

Federici, S. (2004). *Caliban and the witch*. Autonomedia.

Ferguson, S. (2008). Canadian contributions to social reproduction feminism, race and embodied labor. *Race, Gender & Class, 15*(1–2), 42–57.

Fisher, E. (2010). Contemporary technology discourse and the legitimation of capitalism. *European Journal of Social Theory, 13*(2), 229–252.

Food and Agriculture Organization of the United Nations (FAO). (2016). *Global forest products facts and figures 2016*. Retrieved August 30, 2023, from http://www.fao.org/3/I7034EN/i7034en.pdf

Fusaro, M. (2012). Cooperating mercantile networks in the early Mediterranean. *Economic History Review, 65*(2), 701–718.

Glenn, E. N. (2010). *Forced to care: Coercion and caregiving in America*. Harvard University Press.

Harvey, D. (2003). *The new imperialism*. Oxford University Press.

Hirsch, J. (1978). The state apparatus and social reproduction: Elements of a theory of the bourgeois state. In J. Holloway & S. Picciotto (Eds.), *State and capital: A Marxist debate* (pp. 57–107). Edward Arnold London.

Horn, L. C., & Vliegenthart, A. (2010). Corporate governance in Nederland - hoe bedrijven handelswaar werden. In J. Blom (Ed.), *De Kredietcrisis - Een politiek-economisch perspectief* (pp. 59–78). Amsterdam University Press.

Hunter, D. (1978). *Papermaking: The history and technique of an ancient craft*. Dover Publications.

Järvinen, J., Ojala, J., Melander, A., & Lamberg, J.-A. (2012). The evolution of pulp and paper industries in Finland, Sweden, and Norway, 1800–2005. In J.-A. Lamberg, J. Ojala, M. Peltoniemi, & T. Särkkä (Eds.), *The evolution of global paper industry 1800–2050* (pp. 19–47). Springer.

Jensen-Eriksen, N., & Ojala, J. (2015). Tackling market failure or building a cartel? Creation of an investment regulation system in Finnish forest industries. *Enterprise & Society, 16*(3), 521–555.

Jessop, B. (2002). *The future of the capitalist state*. Polity Press.

Jessop, B. (1994). Post-Fordism and the state. In A. Amin (Ed.), *Post-Fordism: A reader* (pp. 251–279). Blackwell.

Jessop, B. (1993). Towards a Schumpeterian workfare state? Preliminary remarks on post-Fordist political economy. *Studies in Political Economy, 40*(1), 7–39.

Jessop, B., & Sum, N.-L. (2006). *Beyond the regulation approach—Putting capitalist economies in their place*. Edward Elgar Publishing.

Jonker, J., & Sluyterman, K. (2000). *Thuis op de Wereldmarkt. Nederlandse handelshuizen door de eeuwen heen*. Sdu Uitgevers.

Koch, M. (2004). European reactions to global challenges: The cases of the Netherlands, the UK, Sweden, Spain, and Germany. *Journal of International & Comparative Social Welfare, 20*(1), 15–22.

Kuitenbrouwer, M., & Schijf, H. (1998). The Dutch colonial business elite at the turn of the century. *Itinerario, 22*(1), 61–86.

MacKenzie, N. G. (2018). Creating market failure: Business-government relations in the British paper-pulp industry, 1950–1980. *Business History Review, 92*(4), 719–741.

Nguyen, D. H. (2023). The political economy of heteronormativity. *Review of Radical Political Economics, 55*(1), 112–131.

Oosthoek, K. J. (2018). State forestry in the Netherlands: From liberalism to nature creation. In K. J. Oosthoek & R. Hölzl (Eds.), *Managing Northern Europe's forests: Histories from the age of improvement to the age of ecology*. Berghahn Books.

Overbeek, H., Van Apeldoorn, B., & Nölke, A. (Eds.). (2007). *The transnational politics of corporate governance regulation*. Routledge.

Panitch, L., & Konings, M. (2009). Myths of neoliberal deregulation. *New Left Review, 57*(May/June), 67–83.

Poulantzas, N. (2000). *State, power, socialism*. Verso.

Särkkä, T. (2021). *Paper and the British empire: The quest for imperial raw materials*. Routledge.

Särkkä, T. (2012). The British paper industry, 1800–2000. In J.-A. Lamberg, J. Ojala, M. Peltoniemi, & T. Särkkä (Eds.), *The evolution of global paper industry 1800–2050* (pp. 167–190). Springer.

Särkkä, T., Gutiérrez-Poch, M., & Kuhlberg, M. (2018). Technological transformation in the global pulp and paper industry: Introduction. In T. Särkkä, M. Gutiérrez-Poch, & M. Kuhlberg (Eds.), *Technological transformation in the global pulp and paper industry 1800–2018: Comparative perspectives* (pp. 1–12). Springer.

Schrauwers, A. (2010). "Regenten"(Gentlemanly) capitalism: Saint-Simonian Technocracy and the emergence of the "Industrialist Great Club" in the mid-nineteenth century Netherlands. *Enterprise and Society, 11*(4), 753–783.

Senchyne, J. (2017). Rags make paper, paper makes money: Material texts and metaphors of capital. *Technology and Culture, 58*(2), 545–555.

Te Velde, H. (2008). The organization of liberty: Dutch liberalism as a case of the history of European constitutional liberalism. *European Journal of Political Theory, 7*(1), 65–79.

Turunen, O. (2012). The paper industry in Germany, 1800–2000. In J.-A. Lamberg, J. Ojala, M. Peltoniemi, & T. Särkkä (Eds.), *The evolution of global paper industry 1800–2050* (pp. 81–99). Springer.

van den Tol, J. (2020). *Lobbying in company*. Brill.

Van Zanden, J. L. (2005). *The economic history of the Netherlands 1914–1995: A small open economy in the 'long' twentieth century*. Routledge.

VNP. (2021). *Jaarverslag 2021–2022*. Retrieved May 30, 2023, from https://www.calameo.com/read/006336446f0bf51830d9a

Wigger, A., & Buch-Hansen, H. (2013). Competition, the global crisis, and alternatives to neoliberal capitalism: A critical engagement with anarchism. *New Political Science, 35*(4), 604–626.

CHAPTER 2

Capital Networks and Early Papermaking

Abstract This chapter explores the emergence of the Dutch paper industry in the phase of Dutch capitalism starting from 1580. Swiftly earning acclaim as the most reputable paper industry in Europe and beyond, the chapter elucidates the distinctions between the primary paper production regions during this period, namely the Zaanstreek and the Veluwe. Offering an intricate analysis of the contrasting organization of employers and laborers, diverse politico-economic conditions and industrial integration, as well as variations in resource availability and pre-industrial innovation in each region, the chapter eloquently highlights the transitional and *non-linear* character of the succession from feudalism to capitalism during that era.

Keywords Dutch capitalism · Merchant-capitalists · Handmade paper · Hollander beater · Veluwe · Zaanstreek

> The story of [the transition from feudalism to capitalism] consists of a real dialectical unity, in which the stalled fragments of capitalist development in one country formed the elements of its further development in the next.
> Brandon, P. (2011, p. 142)

The Dutch paper industry emerged in the phase of Dutch state building around 1580. Soon after, it was the most reputable paper industry in Europe and beyond. The first Dutch paper mill was erected

M. E. Ehrich, *The Dutch Paper Industry from 1580 to the Present*, Palgrave Studies in Economic History,
https://doi.org/10.1007/978-3-031-54324-1_2

in Dordrecht in 1586 by Hans van Aelst, an economic and religious refugee from Antwerp.[1] From then onwards, paper mills started spreading throughout the Northern Lowlands (now the Netherlands). In the first half of the seventeenth century, the papermaking process was similar across all Dutch paper mills and comparable to production processes abroad. Papermaking involved decomposing rags, which was labor-intensive and required special knowledge and training.[2] First, the *lompen* (rags) were sorted, then soaked in vast amounts of water and decomposed into fibers using a *hamerbak* (a heavy wood-pounder with nails). The so-called *schepper* (shoveller) shoveled the resulting pulp into a large container and formed it into a sheet of paper, using copper wire and a sieve. Thereafter, the so-called *koetser* (finisher) dried the paper sheets by bedding them in between felt, pressing the amounted stack and hanging it up. To make the resulting paper product into printable paper, it was soaked in animal-based glue and flattened out afterward. Depending on the degree of flattening mixed with further coatings, the paper reached a higher quality as it became more durable.

The initial rise in paper demand was driven by the establishment of book printing in the second half of the fifteenth century and comparably high rates of literacy throughout the Northern Lowlands at that time.[3] With the fall of Antwerp in 1585 and the subsequent independence of the Seven United Provinces from the Habsburg Empire in 1588, two distinct regions became central to Dutch paper production: The Zaanstreek and the Veluwe. In the Veluwe, more and more peasants gradually shifted their second income stream (their first one being subsistence farming as well as selling over-produce at local markets) to paper production as local demand for paper products grew continuously. The Zaanstreek, on the other hand, attracted extensive capital investments from the merchant-capitalist class fraction, soon constituting a dense capital network and emerging into the main industrial region in the Netherlands—also for papermaking. Cooperation and contractual agreements enabled Zaansian paper producers to implement technological advances, sharing the risks and liabilities of expanding paper production. Furthermore, close relations between state authorities and the merchant-capitalist class fraction ensured the containment of rising workers' revolts.

Dutch paper makers produced different paper qualities in these two regions. In the Veluwe resources were scarce and paper producers had to travel far across the countryside to collect rags, machinery, and other necessities for production. Nonetheless, Veluwian paper makers had access to clean water and, thus, produced the highest quality paper products in terms of low paper contamination.[4] Also Veluwian water-driven mills were reliable in terms of continuous energy availability in comparison to wind-, horsepower- or tidal-driven ones.[5] Paper producers in the Zaanstreek suffered from the irregularity of wind energy for their wind-powered mills to run the wood-pounder and, thus, exhibited lower production capacities than in the Veluwe. Yet, they profited heavily from being embedded in a network of other industries as well as the Amsterdam trade market and the adjacent harbor infrastructure for the cost-effective acquisition of raw material.[6]

Both the thriving dyeing industry and the sugar refinery industry were located in the Zaanstreek. Amsterdam emerged as a key center for the dyeing trade, with its favorable location near waterways facilitating the import of raw materials like dyestuffs, such as indigo, cochineal, and madder, from Dutch colonies across the globe. Particularly vast quantities of dyestuffs were extracted from colonies in Asia. Dutch dyeing techniques and expertise were highly regarded, leading to the establishment of numerous dye houses and workshops throughout the country. The industry primarily focused on the production of vibrant and high-quality dyes for textiles, including wool, silk, and linen. Amsterdam also became a central hub for sugar production and trade. The Dutch extracted raw sugar from their colonies, primarily Brazil and later the Caribbean, heavily relying on the intensive exploitation of enslaved laborers at colonial plantations. These raw sugars were then processed in Dutch refineries into white sugar, which was in high demand all across Europe. Rooted in colonial exploitation, the Zaanstreek industrial region and its associated industries, such as manufacturing, shipping, banking, and trade, continued to thrive.

Zaansian paper makers particularly profited from the cheap availability of and easy accessibility to necessary resources, such as fishery nets, cordage, and textiles, often remnants of the dye and sugar industries. For example, blue shirts, which were popular in the fishery and shipping industries at that time, were the basis for the famous Zaansian blue paper.[7] Other famous Zaansian paper products were known for their brightness and based on high-quality textiles traded at the Amsterdam market.[8] Zaansian paper makers, nonetheless, faced one crucial problem: The absence of clean water. It was a continuous effort to innovate the provision of cleaner water, often in the form of cooperation among multiple paper makers.[9]

The merchant-capitalists were generally supportive of industrial innovation in the paper industry. They recognized the potential for increased profits and economic growth that could come with industrialization. By adopting innovative technologies and mechanized processes, the paper industry could enhance production efficiency, reduce costs, and meet the growing demand for paper. This aligns with the capitalist principles of maximizing profits and capitalizing on market opportunities. The merchant-capitalists also had a vested interest in promoting industrial innovation to maintain their competitive edge in the paper trade. Amsterdam was a key center for paper production and trade, and the merchant-capitalists played a crucial role in facilitating international commerce and establishing global networks. Embracing industrialization would allow them to produce larger quantities of paper, expand their market reach, and compete with other European paper-producing regions. They saw industrial innovation as a means to maintain their dominance and secure their position in the paper industry.

Furthermore, the merchant-capitalists recognized that industrialization in the paper industry could lead to increased specialization and diversification of paper types. This would enable them to cater to different market segments and capitalize on emerging industries, such as publishing, education, and bureaucracy, which relied heavily on paper. Embracing industrial innovation would allow to produce specialized papers more efficiently, opening up new avenues for profit and expanding customer bases. The merchant-capitalists' eagerness to adopt industrial advancements in the paper industry conflicted with their stance toward the dyeing sector. Unlike paper makers, dyers functioned within guilds, thereby upholding a monopoly over the creation and trade of dyed fabrics. Consequently, the merchant-capitalists preferred to maintain the existing artisanal dye production system, which was exceptionally lucrative for them. By overseeing the production process and restricting competition, they could dictate elevated prices for their traded goods. Hence, while the adoption of industrial innovations and machinery in the paper industry promised increased profits for the merchant-capitalists, it could have jeopardized the established dominance of the dyeing industry, as it had the potential to disrupt established power dynamics and lead to heightened competition.

While Zaansian paper grades already stood in competition to the paper produced in the Veluwe, the products of the dyeing industry in the Zaanstreek were actually—without competition—renowned for producing high-quality and vibrant colors, achieved through meticulous manual processes and expertise. Hence, the merchant-capitalists were concerned about the potential impact of industrialization on the quality and reputation of these dyed textiles. Mechanical dyeing methods, though leading to increased efficiency and lower production costs, could severely compromise the quality and intricacy of the dyeing techniques. Additionally, machinery and automation would reduce the demand for skilled artisans and laborers, potentially leading to unemployment and social unrest. So while the merchant-capitalists resisted industrial innovation in the dyeing industry due to economic self-interest, concerns about sinking product quality, and social and economic ramifications, they supported industrial innovation in the paper industry. Here, they saw opportunities for increased profits, market expansion, and rising paper trade.

And they would not be wrong as one particular technological innovation turned the tide between the two core regions of paper production in the Netherlands, enabling a three times faster decomposition of cloth into pulp: The *Hollander* beater.[10] In 1680, a Zaansian paper maker, who remained anonymous, invented this machine to beat cloth in order to retrieve its fibers. In the preceding decade, several paper makers of the Zaanstreek had already used metal blades to decompose cloth and tried to either request a patent for their invention or keep others from receiving such a patent.[11] Since metal blades stained the paper with rust, the Hollander beater was composed of several bronze blades, which rotated on a wooden or metal base. This invention spread throughout the Zaanstreek rather quickly and production facilities expanded rapidly, soon exceeding Veluwian production scales by far.[12] Since the resulting quality of paper produced by the *Hollander* beater was much lower in terms of strength and contaminants than that produced by a wood-pounder, Veluwian paper producers continued to use the less efficient wood-pounder. As a result, Zaansian paper makers grew to symbolize the Dutch paper *nijverheid* (zealousness) by the end of the seventeenth century for which the United Provinces gained worldwide fame.

2.1 Growing Capital Networks

State-industry relations were crucial for paper production, especially in the Zaanstreek. The entrepreneurial Zaansian capitalists belonged to a group of wealthy Baptists, who built a network of trade connections and financial relations. This network of wealthy merchant families also held strong ties with the most powerful of *Staten-generaals* (state generals), *Raadspensionarissen* (grand pensionaries), and *Stadhouders* (governors).[13] Throughout the seventeenth and eighteenth centuries, the Dutch Republic was "a federal state with strong features of bourgeois self-government".[14] The establishment of the first Dutch nation-state was driven by unifying nationalistic themes, such as land and growth, "[...] based in the [proclaimed] superior virtue and corresponding economic wealth of the Dutch people".[15] The Amsterdam-Holland merchants were the main agents behind this nationalistic project[16]; they were either in charge of political offices, so-called *regenten* (patricians), or making sure that their interests were heard in the manifold council meetings of representatives of the government, corporate bodies, and commercial entities. The rise of the Dutch Republic to global hegemonic status was, thus, "a multi-dimensional network of individuals and the institutions in which they wielded power, dynamically creating and recreating themselves and attempting to project their authority and views [...]".[17]

Additionally, the merchant-capitalist class fraction fostered and legitimized the national identity through expanding their colonial activities. The use of Dutch-quality paper was essential for this nationalist project of identity building and colonial exploitation. The merchant-capitalists were the main purchasers of Dutch paper, in order to print maps, books, and propaganda material to 'inform' the Dutch public about their colonial ventures.[18] This propaganda material entailed "[mills] as symbols of modernity and ingenuity, but also as solitary stalwarts in the landscape working for profit in the community by grinding grain, milling paper, or pumping water".[19] The profit-seeking merchant-capitalist class fraction had a great interest in displaying the nexus between national industrial growth and 'colonial trade'[20] in this propaganda material. In turn, the growing commission for propaganda material directly benefited national paper producers. The Dutch paper industry was, thus, closely intertwined with Dutch colonialism and early capitalism, being inter-related forces themselves.[21]

The expansion of the Zaanstreek needs to be understood alongside two further aspects: The introduction of secondary markets and subsequent institutionalization of colonial activities. Initially, secondary financial markets were established for the *Vereenigde Oost-Indische Compagnie* (VOC, United East India Company) and other *rederijen* (shipping companies) to buy and sell bonds and shares.[22] Through issuing transferable shares, liquid capital became easily available, also for investing in Dutch paper production. In the beginning, the *rederijen* comprised a narrow circle of financiers, commonly connected through family bonds and often holding shares of several different companies simultaneously, including shares of paper mills.[23] Through a rigid system called *contracten van correspondentie* (contracts of correspondence), a range of economic activities, whether in terms of national industrial or colonial trade activities, were monopolized and cartelized.[24] In fact, "all governmental and state-chartered corporate spoils in Holland [were divided] among families through a system of rotating offices that generally excluded families outside the oligarchy".[25] Through such a cartel-like division of the market, the merchant-capitalists and state authorities were able to steer and control the majority of national *and* international economic activities.

Additionally, this nexus of powerful elites made sure that property-rights and tax systems were organized in a way that would operate at the cost of everyone else except them. Public debt, for example, which underpinned the Dutch naval and military power, "formed both a secure outlet for capital investment and a source for state-demand".[26] Strictly following the principle of over-taxation, the burden was carried by the urban proletarian class instead of the capital-owners:

> Overtaxation is not an incident, but rather a principle. In Holland, therefore, where this system was first inaugurated, the great patriot, DeWitt, has in his 'Maxims' extolled it as the best system for making the wage labourer submissive, frugal, industrious, and overburdened with labour.[27]

Essentially, the organization of the politico-economic realm was dominated by networks in the form of institutionalized overlaps between political and entrepreneurial-merchant functions as well as dense family bonds, giving the early Dutch state project its title of "state monopoly capitalism".[28]

Consequently, the emergence and expansion of state monopoly capitalism during that time was not solely based on national industrial production, but also driven by colonial exploitative practices, which were facilitated and legitimized through state-sponsored corporate bodies like the VOC. Not mentioned often enough,

> the record of Dutch brutality in enslaving the indigenous peoples (literally and metaphorically) or in depriving them of their means of livelihood, and in using violence to break their resistance to the policies of the Company,

> matched or even surpassed the already abysmal standards established by the crusading Iberians throughout the extra-European world.[29]

It is through such means that, in the seventeenth and early eighteenth century, the Dutch merchant-capitalist class fraction, located in Holland as the political and economic decision center of the Netherlands, expanded internationally by first controlling foreign trade routes and then the majority of the world trade market. Hence, both local industry investments and colonial investments were of vital importance for the accumulation of wealth during the so-called Dutch 'Golden Age'. In fact, inland products comprised the main part of (re-)export, followed by European goods and lastly by colonial goods.[30] Only toward the end of the eighteenth century did colonial goods start outnumbering the (re-)export of inland and European products.[31] Hence, state monopoly capitalism depended on the unique ensemble of national industrial activities *and* colonial exploitation.[32] In essence, a deep interconnection between national production and colonial trafficking in the Dutch Republic existed, contesting the common historical depiction of the Dutch Republic as a merchant-capitalist nation focusing solely on trade and commerce.

To sum up, Zaansian paper production more specifically and national industries more generally need to be understood against the backdrop of colonial exploitation. Not only did Zaansian paper producers rely on second-order trading materials such as cloth and fabric entering through the Amsterdam colonial trade market but more importantly, this as well as other national industries served as profitable investment outlets for the merchant-capitalist class fraction. Contracts of correspondence were pivotal to ensure that the merchant-capitalists' fortunes—made through colonial exploitation—remained in the hands of a few families once they were reinvested. Zaansian paper mills, too, were owned and managed by (share-)capital holders, who reinvested their profits from the slave trade and colonial exploitation under the premise that these would remain in their families for generations.

This nexus of colonial and national industrial capital was enabled by close networks between the merchant-capitalist class fraction, mainly Baptist entrepreneurs, and political authorities. And it is precisely these cartel-like network structures, which are characteristic of state monopoly capitalism. Nicely summed up by Sutton (2015), the aforementioned printed propaganda materials are "artifacts of the dynamic networks and actions of government officials, merchants, and publishers within a complex social structure where visuality complements history, positivism and rationalization, and the concomitant rise of the Dutch nation-state and its capitalistic economy".[33] The nationalist project of forming the Dutch state was based on the relentless expansion of the merchant-capitalist class fraction's profits through investing in colonial and national industrial activities. The symbiotic relationship between colonial trade and the national industry in the United Provinces was shaped by a capital network. This network was characterized by the merchant-capitalist oligarchy, who passed down their capital through generations, tightly bound within their family connections. This practice served to solidify their hegemonic power for decades to come.

2.2 Fire Insurances and Other Cartels

Zaansian paper makers were foremost entrepreneurs, who engendered capitalist logics of accumulating and reinvesting capital in order to expand their market shares with a view on ever-increasing profits. During the late seventeenth century, these paper makers were able to maintain their dominant market position by exporting their products to Spain, Portugal, England, Austria, Sweden, Poland, and Russia.[34] Simultaneously, they invested in foreign paper production, mainly in France. In conjunction with these investments, the expertise and machinery of Dutch paper production, for example, the *Hollander*, was introduced to production sites abroad.[35] Consequently, the first half of the eighteenth century was marked by the widespread replication of paper grades common to Dutch paper by foreign competitors, leading as far as falsifying Dutch watermarks and copying unique recipes for papermaking.[36] In addition, a range of European states took protective measures to nurture their national paper industries during the eighteenth century. The Dutch state, in contrast, lacked a centralized, political authority to provide protective legislation for its national industries; additionally, the dominant class fraction of merchant-capitalists pushed for liberal trade policies.[37] As a result, Dutch industries, including paper manufacturing, were exposed to increased competition, which in turn accelerated the growing density and scale of cartel structures.[38]

To counter heightened competition, Protestant paper mill owners in the Zaanstreek shared rising financial risks of investments through establishing *partenrederijen* (shareholder companies).[39] *Partenrederijen* were copied from similar organizational forms in the shipping industry and divided the ownership of paper mills into *parten* (shares). These legal entities held shareholders responsible for proportionate investments into machinery and stock, the payment of losses, and eligibility for profits.[40] In the context of Zaansian paper mills, these shares were fragmented across a limited network of families.[41] Such forms of sharing risks were only able to secure the stability and longevity of paper mills to a certain extent as expensive machinery, high labor costs, and large scales of production persisted. At that time, the scale of production in a Zaansian paper mill amounted to double the production of a mill in the Veluwe.[42] This fact shall not be mistaken for a generally larger production output in the Zaanstreek than the Veluwe, as mill numbers in the Veluwe still surpassed those of the Zaanstreek by far during the eighteenth century.

To raise their competitiveness toward Veluwian paper production, Zaansian paper makers decided to complement their *partenrederijen* with so-called 'fire insurances' worth up to 240,000 Guilders for its members in 1775.[43] More so than insuring against possible fire damages and loss of production, these paper makers' contracts were similar to the aforementioned contracts of correspondence, in that they intended to stifle competition among Zaansian paper producers. In these contracts, the Zaansian paper makers determined the quantity and quality of production and the sales conditions of paper products. The mill managers' adherence to the verbal, but nevertheless contractually binding agreements made within these 'fire insurances' was controlled through regular inspections of the associated mills.[44] Hence, the label 'fire insurances' was a mere camouflage-tactic as these agreements were outright cartel structures.

Such practices among Zaansian paper producers exemplify the capitalist paradox of, on the one hand, the need for continued accumulation of capital through competition, and, on the other, capitalists seeking to stifle competition, either through cartels or economic concentration.[45] Zaansian paper mills were owned by a network of family-bound shareholders, which led to severe concentration regarding ownership. National competition, in combination with high investment risks, posed a meaningful threat, even in light of the rising demand for paper products. Zaansian paper makers were also threatened by international competitors and decreasing prices for paper products due to growing international supply, mainly from France, England, and Germany. During the eighteenth century, the paper makers' contracts were only one form of cartelization to countervail competitive pressures *and* stay competitive at the same time. The major Dutch paper producers cooperated in further cartels to seek sectoral regulations on product criteria, market share allocations, and price determination.[46] Especially, the cooperative practice among Zaansian paper makers to collectively buy paper mills in order to then shut them down and, thus, reduce local competition, illustrates the paradoxical nature of the capitalist competition regime.

The Zaansian cooperation practices to erode competition stand in stark contrast to the hinterland paper production in the Veluwe, where financial and cooperative ties between the numerous, smaller paper producers were rare.[47] The business of lending money and interest-bearing usury, common among the Zaansian paper makers, was frowned upon as unchristian by the predominantly Catholic farmers in the Veluwe. The different branches of Christianity practiced at that time in the United Provinces were quite influential on the daily practices and work ethics of the paper millers. While Catholic believers under no circumstances engaged in businesses of lending money, usury, or banking, this was a common activity among Zaansian paper makers and bankers, also referred to as Lombardiers. In contrast, in the Southern regions of the United Provinces, including the Veluwe, peasants' "[…] attitude toward capitalism was the old-fashioned Christian attitude which admonished each person to acquire no more worldly possessions than were absolutely necessary".[48] As a consequence, the Veluwian paper mills were auxiliary ventures next to farming. The peasants had to rent the land and property, including the mills, from the local nobles and pay tax to the *stadthouderate* (province) for usage of water and other resources.[49] The family and relatives living and working on the actual farming grounds carried these costs. Their mills were centered on local markets and peasants specialized in producing certain, locally demanded paper products, quite independent of the paper trade at the Amsterdam stock market.

Throughout most of the seventeenth century, the Veluwian farmers did not compete in the production of paper because the demand outreached the supply of paper produced in the region. As cooperation was family-bound in the Veluwe, family members supported each other in repairing machinery or lending labor power when needed. Even though Veluwian paper grades were highly demanded, in times of lowered sales-numbers Veluwian paper makers were to halt their production. This constant state of precarity was further fueled by the landownership structure and unavailability of financial capital. Consequently, the majority of Veluwian paper mills had a rather short lifespan. Only a limited number of Veluwian paper mills are still known at present under their original family names, such as Orges, Pannekoek, Van Delden, Schut, Sanders, and Berends.

To conclude, paper makers in the Veluwe were commonly subjugated to feudally organized land-payments and demand-driven production scales, cooperating in terms of lending machinery and labor power. In contrast, Zaansian paper makers cooperated in close-knit networks, which were inaccessible to peasant paper makers from the Veluwe. Zaansian networks were outright cartels, intended to foster the Zaanstreek paper industry. Furthermore, these forms of networked capital provided a link between a few, rich Zaansian paper makers and the merchant-capitalist class fraction. The latter tolerated these cartels as they themselves sought to expand and increase their profits from colonial and national industrial investments. Nonetheless, as the next section demonstrates, laborers continued to pose a meaningful threat to the profitability of national industrial investment outlets for the merchant-capitalist class fraction.

2.3 Labor Revolts

In the Dutch Republic of 1588, cities gained influence in the government through obtaining seats in the States of Holland and the States General. As a consequence, cities soon started to not only compete for political influence on a national level but also for economic wealth.[50] The cities' new plethora was to maintain public order and acquire skilled workers for their industries. The fact that the Dutch Republic represented a "new alliance between the bourgeoisie and the nobility" helped to safeguard capitalist class relations.[51] Soon, the pressing class struggles were effectively confronted through illegalizing and prosecuting previously successful forms of labor revolts.

At that time, the majority of laborers originated from Flanders and France. They were commonly referred to as *knecht* (servant) and highly influenced by foreign cultures of workers' upheavals.[52] These *knechten* organized to fight against income inequity, hazardous working conditions, and other work-related issues. If conflicts became more structural and could not be solved, the *knechten* commonly refused to work and even left the city altogether to find work elsewhere, called *uitgang*.[53] As *uitgang* was one of the most drastic forms of labor upheavals during the fifteenth and sixteenth centuries, cities were quick to enact new laws, which obliged laborers to possess a paper stating that their previous work contract was terminated under mutual consensus. Workers' abilities to leave unjust working conditions were eliminated, if they wanted to be employed again in other cities of the United Provinces. Additionally, employers started forming networks, in which they comprised lists of so-called blackballed workers, who had left their job in irregular ways. These networks, just as the 'fire insurances', were producer cartels with manifold functions. Not only did they enable technological progress and stifle competition, but they also enabled the merchant-capitalist class to curb labor radicalism and enforce peace.

As the division between capital and labor power was taking on the form of modern capitalism, laborers found new ways of organizing. From 1600 onwards, laborers' meetings became a vital part of the proletarian social life outside of work, in which they discussed important issues and supported each other. As an immediate response, city-states enacted laws, under which such meetings were limited to 12 times a year or banned altogether from taking place in certain areas or settings. In response to this, *court vergadering* (court meetings) became a viable practice to informally solve conflicts between employers and workers.[54] Even though city-states illegalized these practices, employers and workers continued to solve their disputes via such trials. Commonly, the party judged guilty had to pay a fine. In turn, such "fines were sometimes immediately used to buy drinks for all those in attendance, and in other cases they were donated to the city's poor fund".[55] For the employers *and* laborers, these court meetings portrayed a much more viable, fair form of conflict negotiation than the legalized options of official law courts and guilds committees.

By the early eighteenth century, proletarianization was on the rise and most of the described practices of labor organization became limited to the growing city-states, particularly regions like the Zaanstreek (see Table 2.1). Industrial clustering was of utmost importance in the Zaanstreek region, employing roughly 60 percent of the male labor force in the Netherlands.[56] Most of the laborers in Zeeland and Holland were working in the fishing and trading fleet, employed by the VOC, Dutch West India Company (WIC), or navy. All three had bad reputations for belated payments and poor working conditions, with respect to workers' safety and insurance.[57] The working conditions and organizing situations of sailors and industrial workers cannot be compared. Sailors often stayed in their jobs for only a few years before settling down in the colonies. Industrial workers, on the other hand, were increasingly forced to not only stay within their profession but also within the same country or city due to the aforementioned decree of employers' appraisal.

Table 2.1 Comparing Dutch paper production regions, sixteenth to eighteenth centuries

	Zaanstreek	*Veluwe*
Technology	• close proximity to resources • clean water scarcity • discontinuity of wind-powered mills • cost-intensive innovations • implementation of *Hollander* to raise efficiency	• cloth available at further distance and higher prices • clean water and wind-powered mills • handmade paper production based on traditional hamerbak
State-industry relations	• networks of merchant-capitalist class fraction and state authorities • close ties between colonial and national industries	• side-stream earning for peasants • tax payments to local authorities
Competition and cooperation	• contracts of correspondence • cartel structures	• independent of other industries and trade markets • focus on local demand in rural markets • no usury or lending money • local support between families
Labor-capital relations	• rising proletarianization • workers' organizations and strikes	• patriarchal family structures • no workers' organization

To subtend the often conflicting relationships with their direct supervisor as well as the owner of the production facility, paper mill workers (usually around 25 laborers per paper mill) started organizing themselves.[58] The provincial government labeled such organizing as complots, observed them carefully, and ultimately managed to efficiently suppress them. As "the government could rely on a loyal standing army [to maintain] public order" workers' organizations were rarely influential, except in the textile industry.[59] Not surprisingly, the government was highly selective with respect to the forms of industrial organizing it tolerated or even supported and which ones it illegalized. While employers' cartel structures and organizing to list 'unruly' workers, among others, were welcomed by state officials, various forms of workers' organizations were illegalized as soon as they gained popularity. Solidarity between workers was essential to oppose political authorities and the owners of production alike. Protests of the eighteenth century were driven by a code of honor among the workers, which demeaned anyone who continued to attend their work during a strike. Additionally, protests and related fights between workers and employers were to happen in public, never in the private space of a house or a workshop.

The situation of organized labor was very different in the countryside. While the agrarian sectors near the industrialized regions of Holland (including Friesland and later Groningen) were already commercialized, peasants throughout the regions of Gelderland, Drenthe, and Overijssel were not part of the urban communities of workers. The Veluwian paper production as an industrial side-stream of feudally organized agriculture was based on patriarchal[60] family structures. One of the men in the household commonly organized production, determined every family members' involvement and duties, and actively engaged in the production process. Usually between four and six people worked at these mills, of which one-third were children.[61] They received little to no payment. The division of capital and labor in the Veluwe also manifested in the rent payments toward the land-owning nobles and the dependence of peasants on urban merchants to sell their end-products. Hence, Veluwian paper production could be considered as an example of the Industrious Revolution, a significant shift in how people approached work and leisure in Europe during the seventeenth and eighteenth centuries, resulting in increased productivity and economic growth.[62] Veluwian paper production was ultimately a family business, relying on local markets and resources to secure the profitability of the papermaking side-stream. Zaansian paper production, on the other hand, could be considered an example of early industrialization. Zaansian paper makers were the foremost investors of family-bound money made through colonial exploitation. Close ties to the merchant-capitalist class fraction allowed for costly, technological innovations to be developed and implemented.

The analyzed differences between the Zaansian and Veluwian paper production concerning each of the four dimensions during Dutch capitalism are striking (see Table 2.1). Regional differences in the organizing of employers and laborers, the different politico-economic circumstances and industrial embeddedness of each region, and the differences in resource availability and pre-industrial innovation of the Zaanstreek and the Veluwe beautifully illustrate the transitional and *non-linear* character of the succession from feudalism to capitalism. In the words of Federici (2004), "[t]he concept of 'transition' then, helps us to think of a prolonged process of change and of societies in which capitalist accumulation *coexisted* with political formations not yet predominantly capitalistic".[63] Consequently, feudally organized paper production in the Veluwe coexisted alongside more capitalistic organized paper production in the Zaanstreek.

Ultimately, the phase of economic growth in the Dutch Republic, which reached its zenith in the 1650s, was followed by a general decline in labor power, prices, production, and profit. Starting in 1670, decreasing Dutch colonial power and economic stagnation took its toll on both paper regions, the Zaanstreek and the Veluwe. These struggles were interlinked with the four Anglo-Dutch wars and

> [...] the seventeenth century crisis, which reduced prices of staple exports and demand for luxury goods [and dragged] the other sectors of Dutch production down. As a result, Dutch élites withdrew into 'extra-economic' strategies and investment in politically constituted property such as public office.[64]

Many wealthy merchant families, constituting the majority of capital investors, no longer saw national industrial investments as secure and viable outlets. Instead, they started investing into foreign financial markets. The rule of William III, the king of Ireland, England, and the United Provinces, added to the economic devastation and political isolation of the United Provinces at that time. As he granted the Royal Charter in 1694 to the Bank of England, the central role of Amsterdam as the world capital city quickly diminished.[65] In 1702, William III died and left behind a financially exhausted Dutch Republic, the world trade center having shifted to London.[66] The flight of capital, monetary destabilization, and the inability to compensate for these developments through emergency taxation measures intensified under Napoleon's rule from 1795 to 1813.[67] International trade diminished vastly and the Dutch paper industry shrunk alongside other national industries. While resource prices rose, prices of paper products declined. Lastly, the ban on importing rags and high taxation on the national usage of rags led to the shutdown of a vast number of paper mills in the Veluwe as well as the Zaanstreek.[68] In the following phase of monarchic liberalism, these circumstances demanded new forms of networks to bolster and restructure Dutch papermaking.

Notes

1. Linssen (1988, p. 14).
2. Voorn (1975, pp. 17–19).
3. Bouwens (2004, p. 21).
4. De Vries (1957, p. 13).
5. De Wit (1990, p. 8).
6. Ibid., p. 9.
7. De Vries (1957, p. 15).
8. Ibid., p. 52.
9. De Wit (1990, p. 13).
10. De Vries (1957, p. 16).
11. Stichting Papiergeschiedenis Zaanstreek de Hollander (n.d.).
12. De Vries (1957, p. 51).
13. Brandon (2011, pp. 120–123, 131, 138) and Lachmann (2000, p. 162).
14. Brandon (2015, p. 6).
15. Sutton (2015, p. 130).
16. E.g. Adams (1994, p. 516).
17. Sutton (2015, p. 23).
18. Dekker (1990, 368).
19. Sutton (2015, p. 53).
20. For lack of a better term I use 'colonial trade' as a description of the relations between conquering and conquered populations, in which the conquered populations are "forced to engage in structurally dependent relationship[s] with the [conquering populations] by engaging in *unfair and exploitative trade* relations". Kieh and Wong (2014, p. 16, *italics added*).
21. Anievas and Nisancioglu (2015).
22. Gelderblom and Jonker (2004, p. 643).
23. Davids (2006, p. 560).
24. Adams (1994, p. 516).
25. Lachmann (2000, p. 164).
26. Brandon (2011, p. 116).
27. Marx (1867, p. 534).
28. Arrighi (2010, p. 181).

29. Ibid., p. 159.
30. De Vries and Van der Woude (1997, p. 499).
31. Ibid.
32. Selwyn (2013) and Brandon (2011).
33. Sutton (2015, p. 5).
34. Bouwens (2004, p. 31).
35. De Vries (1957, p. 34).
36. Bouwens (2012, p. 194).
37. Findlay and O'Rourke (2007, p. 395).
38. Bouwens (2004, p. 32).
39. Bouwens (2012, p. 193).
40. Gelderblom and Jonker (2004, p. 645).
41. Davids (2006, p. 560).
42. De Vries (1957, p. 310).
43. Ibid., p. 121.
44. Bouwens (2004, p. 349) and Davids (2006, p. 564).
45. Wigger and Buch-Hansen (2013, pp. 607–609).
46. Bouwens (2004, pp. 33–38) and Davids (2006, p. 22).
47. Kokke (1961, pp. 13–15).
48. Hyma (1938, p. 327).
49. Kokke (1961, p. 12).
50. Dekker (1990, p. 398).
51. Federici (2004, p. 49).
52. Van Vree (2008, p. 169).
53. Dekker (1990, p. 391) and Van Vree (2008, p. 7).
54. Dekker (1990, p. 393).
55. Ibid, p. 394.
56. Bouwens (2004, p. 26).
57. Dekker (1990, p. 406).
58. Bouwens (2004, p. 28).
59. Dekker (1990, p. 387) and Van Vree (2008, p. 169).
60. Patriarchy, being a pivotal organizing principle for and under capitalism, was, of course, also common in the more industrialized and commercialized regions of the Netherlands (for further exploration see Adams [1994, 2005] and Hartsock [1983, pp. 291–293]). The patriarchal custom of passing on family capital and ownership through arranged marriages was closely related to the system of *Contracts of Correspondence*. This is how, for example, Van Gelder Zonen was established: "[…] Pieter Smidt van Gelder decided to

become a manufacturer early on in his life. He started working for Maarten Jansz Schouten, the only associate of the company Maarten Schouten & Co. and since 1774 the owner of the paper mill 'De Eendragt' te Wormer. On 30st November 1783, Pieter Smidt van Gelder was matrimonially arranged to marry the only daughter of his employer Dieuwertje Schouten. A year later, his father in law made him shareholder of the company. The accession of Pieter Smidt van Gelder into the managing board of Schouten & Co. was the starting point for the well-known family firm van Gelder (en Zonen), which was passed on for generations" (Sanders, 1995, p. 280, own translation).

61. De Wit (1990, p. 15).
62. Vries (2008).
63. Federici (2004, p. 62, *italics added*).
64. Brandon (2011, p. 135).
65. De Jong (2011, p. 51).
66. Anievas and Nisancioglu (2015, pp. 196–198, 261–262).
67. Van Zanden and Van Riel (2004, pp. 52–52).
68. Bouwens (2004, p. 35).

References

Adams, J. (2005). The rule of the father: Patriarchy and patrimonialism in early modern Europe. In C. Camic, P. S. Gorski, & D. M. Trubek (Eds.), *Max Weber's economy and society: A critical companion* (pp. 237–266). Stanford University Press.

Adams, J. (1994). The familial state: Elite family practices and state-making in the early modern Netherlands. *Theory and Society, 23*(4), 505–539.

Anievas, A., & Nisancioglu, K. (2015). *How the West came to rule*. Pluto Press.

Arrighi, G. (2010). *The long twentieth century: Money, power and the origins of our times*. Verso.

Bouwens, B. (2012). The paper and board industry in the Netherlands, 1800–2000. In J. A. Lamberg, J. Ojala, M. Peltoniemi, & T. Särkkä (Eds.), *The evolution of global paper industry 1800–2050* (pp. 191–209). Springer.

Bouwens, B. (2004). *Op papier gesteld: de geschiedenis van de Nederlandse papier- en kartonindustrie in de twintigste eeuw*. Boom.

Brandon, P. (2015). *War, capital, and the Dutch State* (1588–1795). Brill.

Brandon, P. (2011). Marxism and the 'Dutch miracle': The Dutch Republic and the transition-debate. *Historical Materialism, 19*(3), 106–146.

Davids, K. (2006). The transformation of an old industrial district: Firms, family, and mutuality in the Zaanstreek between 1840 and 1920. *Enterprise and Society, 7*(3), 550–580.

Dekker, R. (1990). Labor conflicts and working-class culture in early modern Holland. *International Review of Social History, 35*(3), 377–420.

De Jong, J. (2011). The Dutch golden age and globalization: History and heritage, legacies and contestations. *Global Citizenship, 27*(1), 46–67.

De Vries, B. W. (1957). *De Nederlandse papiernijverheid in de negentiende eeuw*. M. Nijhoff.

De Vries, J., & Van der Woude, A. (1997). *The first modern economy: Success, failure, and perseverance of the Dutch economy, 1500–1815*. Cambridge University Press.

De Wit, O. (1990). *Papierfabricage in nederland in de negentiende eeuw van molen naar fabriek*. Stichting Historie der Techniek.

Federici, S. (2004). *Caliban and the witch*. Autonomedia.

Findlay, R., & O'Rourke, K. H. (2007). *Power and plenty: Trade, war, and the world economy in the second millennium*. Princeton University Press.

Gelderblom, O., & Jonker, J. (2004). Completing a financial revolution: The finance of the Dutch East India trade and the rise of the Amsterdam capital market, 1595–1612. *The Journal of Economic History, 64*(3), 641–672.

Hartsock, N. C. M. (1983). The feminist standpoint: Developing the ground for a specifically feminist historical materialism. In S. Harding & M. B. Hintikka (Eds.), *Discovering reality* (Vol. 161, pp. 283–310). Springer.

Hyma, A. (1938). Calvinism and capitalism in the Netherlands, 1555–1700. *The Journal of Modern History, 10*(3), 321–343.

Kieh, G. K., & Wong, P. N. (2014). The small powers in world politics. *African and Asian Studies, 13*(1–2), 13–32.

Kokke, C. T. (1961). *De Veluwse papiermolen*. Rijksmuseum voor Volkskunde "Het Nederlands Openluchtmuseum".

Lachmann, R. (2000). *Capitalists in spite of themselves: Elite conflict and economic transitions in early modern Europe*. Oxford University Press.

Linssen, G. C. P. (1988). *Papier in Nijmegen: Voorgeschiedenis en ontwikkeling van KPN Nijmegen b.v. ter Gelegenheid van het tachtigjarig Bestaan in 1988*. Janssen Print.

Marx, K. (1867). *Capital, volume I*. Penguin/New Left Review.

Sanders, G. P. (1995). Portret van een niet langer onbekende man: Pieter Hendrik van Gelder, het gezicht van de penningverzameling in het Rijksmuseum. *Bulletin van Het Rijksmuseum, 4*(4), 280–307.

Selwyn, B. (2013). Karl Marx, class struggle and labor-centered development. *Global Labor Journal, 4*(1), 48–70.

Stichting Papiergeschiedenis Zaanstreek de Hollander. (n.d.). *Doel van 'DE HOLLANDER'*. Retrieved May 31, 2023, from http://zaansepapiergeschiedenis.nl/stichting/doel-van-de-hollander/

Sutton, E. A. (2015). *Capitalism and cartography in the Dutch golden age*. University of Chicago Press.

Van Vree, W. (2008). *Nederland als Vergaderland*. Amsterdam University Press.

Van Zanden, J. L., & Van Riel, A. (2004). *The strictures of inheritance: The Dutch economy in the nineteenth century*. Princeton University Press.

Voorn, H. (1975). *Papierfabricage in de eerste helft van de negentiende eeuw*. Vereniging "Vrienden van de Koninklijke Bibliotheek".

Vries, J. (2008). *The industrious revolution: Consumer behavior and the household economy, 1650 to the present*. Cambridge University Press.

Wigger, A., & Buch-Hansen, H. (2013). Competition, the global crisis, and alternatives to neoliberal capitalism: A critical engagement with anarchism. *New Political Science, 35*(4), 604–626.

CHAPTER 3

Building Paper Industria

Abstract This chapter delves into the evolution of the Dutch paper industry during the era of Dutch monarchic liberalism, spanning the years between 1815 and 1914. In the early nineteenth century, the industry faced a period of stagnation attributed to the widespread adoption of steam-powered mills for chemically produced paper throughout Europe and Northern America. While this marked a significant transformation for the Dutch paper industry, transitioning from small-scale, handmade, peasant-based paper manufacturing to the dominance of a large-scale, corporate sector, the establishment of Industria served to sustain Dutch paper production without allowing it to match the industrialization levels seen in neighboring countries.

Keywords Dutch monarchic liberalism · Industria · Steam-run paper mills · Gentlemanly capitalists · Corporatization

According to Schrauwers (2010, p. 753) the term *Industria* describes the nineteenth century "majestic home of the Industrialist Great Club", which is made up of aristocratic financiers turned Saint-Simonian-inspired industrialists under the leadership of their Merchant King.

M. E. Ehrich, *The Dutch Paper Industry from 1580 to the Present*, Palgrave Studies in Economic History,
https://doi.org/10.1007/978-3-031-54324-1_3

> In the course of centralization, a whole host of individual capitals disappear, absorbed by others, while yet others fuse together by merger or consolidation. The centralization of capital is therefore a violent form of competition.
>
> Aglietta (2000, p. 219)

During the beginning of the nineteenth century, the Dutch paper industry suffered from industrial stagnation due to the introduction and rapid spread of steam-powered mills for chemically produced paper all over Europe and Northern America.[1] This powerful innovation was too capital-intensive during the early and mid-nineteenth century to be implemented efficiently in the Netherlands. Machines had to run continuously in order to be profitable. Yet, Dutch paper market outlets could not accommodate such a rising scale of production. Thus, Dutch paper producers had a strong interest in cooperating closely with the government to establish national subsidiaries for paper mills and the legalization of cartels, ultimately aiming to restructure and centralize the Dutch paper industry. These demands were met by state authorities with political agendas of enacting favorable policies or even nationalizing parts of the industry.[2] Hence, networks in terms of close ties between industry agents and state authorities as well as among the paper industrialists themselves were pivotal to successfully meet the challenges of the early and mid-nineteenth century.[3]

From 1840 onwards, close cooperation between the monarch and the industrialist class fraction ensured foreign capital investment interests in Dutch industries and, thus, growing industrialization. Networks, in the form of cartels, were just as pivotal to re-establishing international competitive advantages as were other forms of networks between, for example, the industrialists and the monarchy to secure national subsidiaries.[4] The new monarchy was efficiently co-opted by and itself co-opting the changing patrician and merchant-capitalist class fractions. This interaction enabled the corporatization of the Dutch paper industry under the unique construction of networked capital at that time. As a

[1] De Wit (1990, pp. 20–23).

[2] De Vries (1957, p. 175).

[3] E.g. Jonker (1996).

[4] De Vries (1957, p. 285).

result, the way was paved for the gentlemanly capitalists to build their very own Industria.

More generally, these industrial developments took place in a politico-economic environment, which is referred to as monarchic liberalism.[5] In this capitalist phase, from 1815 until 1914, the Netherlands underwent a state-orchestrated liberalization under the rule of the monarchs. While the networks of merchant regent families of the so-called 'Golden Age' continued to profit under the reign of William I of the Netherlands, they were gradually replaced by the growing class fraction of gentlemanly capitalists during the liberal revolution of 1840 with the emergence of managerial science and technocracy, and the reign of William II of the Netherlands.[6] These gentlemanly capitalists were a new tier: They emerged out of the patrician (*regenten*) families and the merchant-capitalist class fraction to form a "close-knit elite network [...]—the Industrialist Great Club—[and] built its majestic home, Industria, on the former site of the world's first stock exchange".[7] The gentlemanly capitalists managed to use the mid-nineteenth century conjuncture of the succession from William I to William II and the economic crisis to change the power dynamics among the dominant class fractions. They formed "interlocks that transformed nominally independent corporations into coordinated networks"—nothing less than the modern corporation.[8]

National industries were saved from subordination to monarchic power and merchant class interests by subjecting trade, the finance sector, and national industrial production to the newly dominant class fraction: The gentlemanly capitalists. Essentially still resting on a dense network of families (just like their patrician and merchant-capitalist predecessors), the gentlemanly capitalists steered the nationalist project and accumulated capital by nothing less than expropriating the position of William I, nicknamed *the King-Merchant*, with a focus on promoting industrial growth, ultimately exerting control over the entire Dutch state.[9]

Nineteenth-century Dutch liberalism designates a historically specific conjuncture of a hierarchical order of class fractions, which fostered

[5] Martins (2013, p. 148) and Milanovi (2013).

[6] De Vries (1957, p. 175) and Davids (2006).

[7] Schrauwers (2010, pp. 754, 777, *hyphens added*).

[8] Ibid., p. 755.

[9] Ibid., p. 778.

corporate governmentality as the driving force behind the establishment of the modern Dutch nation-state, the restructuring of its national and colonial industries, and the renewal of its economy.[10] Accordingly, Dutch monarchic liberalism was an "[...] apologia of individualistic capitalism [*and*] freedom in restraint".[11] Therefore, the Dutch liberalists of the second half of the nineteenth century and their constitutional demands have to be viewed in direct relation to the co-optation of Dutch monarchy under William I and II.

3.1 The Rise of the Gentlemanly Capitalists

During William I's main politico-economic focus, namely the pursuit of wealth creation through expanding colonial trade, the Northern Netherlands[12] was marked by industrial stagnation in the first half of the nineteenth century. William I directed significant efforts toward reorganizing the commerce sector, prioritizing the interests of the Northern merchant class, even though this fraction's influence was waning.[13] Despite the decreasing importance of the Amsterdam stock market in the global trade landscape and the VOC's bankruptcy in 1799, resulting in a reduced Dutch colonial presence, William I remained enthusiastic about colonial trade. Instead, these events served as a catalyst, driving his interest even further. Subsequently, national industries were barely promoted and the general economic infrastructure of the country could not catch up with industrialization processes abroad.

In alignment with his concerns for commerce, William I followed strategies of "corporate governmentality", namely the delegation of sovereignty to chartered corporations at the beginning of the nineteenth century in the Netherlands.[14] Instead of simply extending absolutist state repression, William I sought to connect with various class fractions, ensuring a new order of the national economy, while at the same time

[10] E.g. Jonker (1996).

[11] Te Velde (2008, p. 67).

[12] From 1815 until 1839, the United Kingdom of the Netherlands consisted of the former United Provinces (Northern Netherlands), Belgium, and Luxembourg (Southern Netherlands). With the declaration of independence of Belgium and Luxembourg in 1839, the Kingdom was split and the present-day borders of the Netherlands were declared legal.

[13] Schrauwers (2010, p. 375).

[14] Ibid., pp. 376–377.

shaping workers' subjectivities under a newly transformed management science. Concerning national industries, William I installed the so-called *Funds for Industry*, which distributed subsidies to "[…] lend support to especially those sectors of the national industry that can not be provided with adequate protection without raising tariffs to such levels that these could have a disadvantageous influence on commerce".[15] Concerning the restructuring of colonial exploitation, he established new commercial companies that replaced the bankrupted VOC, such as the *Nederlandsche Handel-Maatschappij* (NHM, The Netherlands Trading Society). By shareholding in these corporations, William I and the Northern merchant class earned up to 39 million florins per year.[16] While the label for Dutch colonial exploitation changed from VOC to NHM, its forces and results were similar to those of the first phase of Dutch capitalism, generating "'indigenous' revolts against forced labor and starvation".[17] These were directed against institutions such as the 'culture system', which determined certain amounts of colonial production for export, while pocketing the total revenue for the NHM and, thus, William I and the Northern merchant class. The vast majority of this capitalist slavery money was reinvested into financial speculation rather than national industrial developments, in turn manifesting the general trend of national industrial stagnation that marked the first half of nineteenth-century Netherlands.

Starting with the reign of William II, but especially from the liberal revolution of 1840 onwards, government authorities implemented free-trade policies.[18] At the same time, the gentlemanly capitalist class fraction began to successfully implement their particular interests as general interests at the level of the state. Their successes were due to newly developing forms of cooperation, such as the establishment of the Chamber of Commerce in 1843. "Founded by the municipality in Zaandam [it] brought together representatives from different trades and industries […] to pressure government authorities to take measures for improving the Zaanstreek's infrastructure".[19] The government started investing public money into building a national infrastructure and revolutionizing capital

[15] As cited in: Van Zanden and Van Riel (2004, p. 94).

[16] Bourrinet (2016, p. 16).

[17] Ibid.

[18] De Vries (1957, pp. 295–296).

[19] Davids (2006, p. 574).

markets, transport, and communication in order to lower "the costs that entrepreneurs had to make to compete on international markets".[20]

Many more examples show, how well the interests of the gentlemanly capitalist class fraction and the liberals in parliament aligned. In 1869, the government abolished the tax on newsprint to cause a boom in paper demand and thereby combat the paper industry's slow growth and stagnating investments.[21] This strategy yielded an immediate effect on investors' interests in the Dutch paper industry. Additionally, the inaccessibility of high-quality Dutch *lompen* outside of the Netherlands due to a national decree on restricting the export thereof and the apparent lack of industrialization within the Dutch paper industry led foreign financiers to seek profitable chances for investment.[22] Soon these developments stimulated four Belgian bankers and paper makers to found the *Koninklijke Nederlandse Papierfabriek* (KNP, The Royal Dutch Paper Factory).

Other state strategies also fostered investors' interests in Dutch industries. The restructuring of the monetary system was a long-awaited project, driven by the interests of the financial class fraction to allow for an "[…] efficient mediation by brokers and cashiers, [so that] almost everyone with surplus money or a (temporary) shortage could participate on this market".[23] Part of these financial reforms was a restructuring of the first national bank's management, *the Nederlandsche Bank* (DNB, The Dutch Bank). Already founded in 1814, William I had forced the bank to support government expenditures for centuries, thereby greatly diminishing the financial class fraction's trust in this organization.[24] Nevertheless, the bank's reorganization, the restructuring of the monetary system more generally, and the continuing growth of secondary financial markets yielded a rise in national stock and foreign security investments and assured continuous investments into national industries by the gentlemanly capitalists as well as the financial class fraction.[25]

The above delineations illustrate the ability of the gentlemanly capitalist class fraction to establish joint forces, first with the monarchs and

[20] Van Zanden and Van Riel (2004, p. 218).

[21] Pfann and Van Kranenburg (2003, p. 63).

[22] De Groot (2001, p. 288).

[23] Van Zanden and Van Riel (2004, p. 213).

[24] Ibid., p. 163.

[25] Ibid., p. 213.

later with the liberals in parliament, in order to restructure national industries through state policies and interventions. Consequently, state-industry relations during Dutch monarchic liberalism are not marked by laissez-faire strategies and little to no protection for national industries, but by networked capital among the dominant class fractions as shown in the case of the Dutch paper industry.

3.2 Producing Paper of Endless Length

The phase of monarchic liberalism is marked by two major changes within the paper industry: The shift from handmade paper to mechanically produced paper and the change from cloth to cellulose fibers. At the beginning of the nineteenth century, the blueprint for a machine that mechanically produced paper was created. The inventor, Nicholas-Louis Robert, and "his master, Francois Didot, grew impatient with the irascibility and ill temper of the workers, and it was this constant wrangling and discord, […] that gave Robert the impetus to devise a paper-making machine".[26] The Fourdrinier brothers, all engineers, perfected Robert's machine soon after it was patented in 1798. The machine turned vegetable fibers, which were wetted beforehand, into a web of dry paper by passing them over a large number of rollers, removing the water through suction and drainage. The matted and intertwined fibers are called felt, which was further dried by being pressed by steam-heated cylinders. Essentially, the machine imitated the production process of handmade paper. Yet, using mechanization—and in contrast to artisan production—it became possible to produce paper of endless length, the only restriction being the width of the machine that would determine the width of the paper. Even "a child" was able to operate such a machine; there was no longer any need for experts (or, as Robert would have called them, irascible and ill-tempered workers), who would know the detailed and demanding production process of handmade paper.[27] Soon the blueprints for the machine were introduced in the United Kingdom, Russia, and the United States of America, where they were perfected by engineers and put to use by solvent paper makers to make production more profitable.

[26] Hunter (1978, p. 343).

[27] Ibid., p. 347.

International competitive pressures grew immensely for the Dutch paper industry due to the rise of mechanical paper production in the United Kingdom, France, Germany, the United States of America, Russia, and Scandinavia from 1780 onwards.[28] The industry's international competitors produced large quantities of paper at ever-lower prices, having coal-fueled steam machines fully implemented and wood fiber readily available. Due to the high costs of adapting papermaking machines and a mentality that favored handmade paper production processes as well as handmade paper products, the Netherlands with its small- and medium-sized paper millers stayed far behind its international competitors.[29]

Also on the national level, the competitiveness of Zaansian paper production decreased immensely around that time. Smaller Veluwian paper mills had always outnumbered the larger Zaansian ones. This gap increased steadily over time, even though both regions show decreasing numbers of paper mills (see Fig. 3.1). While in the early nineteenth century Zaansian paper production output was double as high as the Veluwian one, the roles were reversed by 1854.[30] The reasons for the immense decline in the number of Zaansian as well as Veluwian paper mills during the nineteenth century are threefold. First, the combination of high protectionism abroad and low protectionism in the Netherlands had for too long favored the interests of the merchant-capitalist class fraction.[31] In fact, during the end of the eighteenth century, protectionism was a general trend throughout Europe to safeguard national industries, virtually closing down main consumer markets for Dutch paper producers.[32] In addition, with only a small success to convey their interests on a national level, Dutch paper producers continued to suffer from low import taxes for paper, which played into the hands of Dutch paper traders as well as foreign paper producers.[33] Second, the uniqueness of Dutch paper regarding its durability and whiteness became outdated as soon as chemical paper strengthening and brightening techniques

[28] De Wit (1990, p. 24).

[29] Hunter (1978, p. 356).

[30] De Wit (1990, p. 19).

[31] Brandon (2011, p. 141).

[32] De Vries (1957, p. 285).

[33] De Wit (1990, p. 20).

were implemented in other paper-producing countries.[34] Third, the most pressing issue, which led to the decline of the Dutch paper industry, was the cost of implementing steam-run paper machines.[35]

To revive the Zaansian paper production and its (inter-)national competitiveness, William I and one of the biggest Zaansian paper producers of that time, Van Gelder Schouten & C., joined forces to implement the first paper machine in the Netherlands.[36] In fact, van Gelder bought the paper mill Het Fortuin te Zaandijk in 1837 and the minister for industry advised to order a steam-run paper machine through the *Rotterdamse Stoomboot Maatschappij* (RSM, The Rotterdam Steamship Company), established and partially owned by William I.[37] Yet, a series of difficulties led to the economic failing of this networked effort, namely the belated delivery and malfunctioning of the machine as well as the costliness of personnel able to operate such machines.[38] When William I

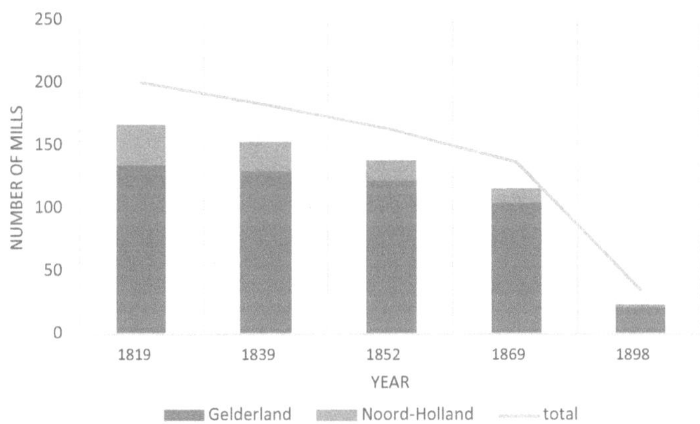

Fig. 3.1 Number of paper mills in the Netherlands during the nineteenth century (*Source* Own calculation based on De Vries [1957, p. 276] and De Wit [1990, p. 19])

[34] Ibid., p. 21.

[35] De Vries (1957, p. 235).

[36] De Wit (1990, p. 25).

[37] Ibid.

[38] De Wit (1990, p. 25).

granted van Gelder the first license to install a steam-run paper machine, he did not tax van Gelder the effective amount of six percent on the import of the machine but demanded the machine to be fueled with peat instead of coal. Overtly, the reason for William I's demand was complaints of neighboring white paper producers, who feared that coal fumes would decrease the whiteness of their paper; covertly, the independence of Belgium ripped William I's kingdom of its coal resources, leaving only the much costlier and less economic exploitation of peat in Dutch territory.[39] Although the king showed a keen interest in Dutch industries utilizing domestically sourced peat rather than foreign coal, relying on peat as a fuel source ultimately led to a significant rise in the machine's economic inefficiency, reaching a point that was financially unsustainable. Though this first adaption of a steam-run paper machine for Dutch production failed, other paper producers followed. Similarly, their attempts to transition from handmade to mechanically produced paper were also unsuccessful from an economic standpoint.[40]

By 1870, the implementation of steam-run paper machines finally took a grip on the Dutch paper industry; yet only capital-strong companies were able to successfully switch from handmade to mechanical paper production.[41] Machines needed to run at full capacity to be profitable and the production output continued to exceed national paper demand by far. Thus, the two big paper producers of that time, VGZ and KNP, once more turned to the government for support. This time with the intention to force a nationwide switch from handmade to mechanically produced paper, in order to decrease national competition. To do so, VGZ and KNP filed a complaint with the Ministry of Home Affairs in 1879, against the cartel structures of the smaller paper producers.[42]

This cartel structure, called Marten Orges, was an association of the remaining 30 producers of handmade paper, mainly located throughout the Veluwe, to agree on selling and buying prices.[43] Right after its establishment in 1872, the handmade paper producers successfully put pressure

[39] Ibid., p. 26.

[40] Ibid., p. 27.

[41] Davids (2006, p. 557).

[42] Bouwens (2004, p. 47).

[43] Ibid., p. 46.

on national and colonial governments regarding paper import prices.[44] In 1879, the national printing press renewed an agreement in direct cooperation with the association Marten Orges, which was based on an original advice of the Ministry of Home Affairs in 1848 for all governmental organizations to only make use of handmade paper to boost the industry.[45] This collided with the complaint filed by VGZ and KNP, which ultimately led the Minister for Home Affairs to denounce Marten Orges as a form of unauthorized collusion, demanding from all governmental organizations to only make use of mechanically produced paper from now on. Thus, KNP and VGZ succeeded in imposing the exclusive use of mechanically produced paper by the state, including the national printing press. In doing so, all other paper mill owners were forced to either borrow money to mechanize their production or shut down. By 1890, the association Marten Orges was suspended and all Dutch paper mills had either mechanized (except van Houtum, Tzn Renkum, and Schut) or shut down their production.[46]

Shortly after the introduction of the papermaking machine across different countries as well as in the Netherlands, a search for less costly and more durable fibers began, as the mechanization allowed for the exploitation of new raw materials. From 1800 onwards, the Dutch government eagerly promoted the nationwide formation of different sectoral commissions, which were intended to spread relevant knowledge and information about production techniques and promote cooperation.[47] The search for more efficient raw materials was taken up by, among others, *Beerta*, a sub-division of the *Genootschap ter bevordering der Nijverheid te Onderdendam* (Society for the Advancement of Industry in Onderdendam), which was an association and cartel-like structure founded in 1837 by the *here(n)boeren* (gentlemanly farmers) as well as state officials, merchants and industrialists.[48] Gentlemanly farmers, in opposition to peasant farmers, were an elite of landowners, originating in the middle ages, who engaged in more commercialized agricultural ventures and held local political offices. The commission *Beerta*

[44] De Wit (1990, p. 37).

[45] Ibid.

[46] Bouwens (2004, p. 47).

[47] Knippenberg and Pater (1990, p. 101).

[48] Archieven (2015).

researched the processing possibilities of straw in order to make the *Dollardpolder*[49] straw production more profitable for farmers.[50] The technical feasibility and introduction of straw-board production quickly generated new capital investments.[51] At the end of the nineteenth century, the share of straw (including esparto and old paper) in Dutch paper production increased, while the use of cloth as a raw material steadily decreased (see Fig. 3.2).

Between 1870 and 1880 about 20 straw-based paper production sites opened throughout the region of Groningen.[52] While some of these were speculative investment outlets for entrepreneurial industrialists, others were cooperatives of farmers, who were eager to countervail the growing hostility of market pressure and low-cost selling arrangements for their overproduction of straw.[53] Even though the first of these cooperatives was not successful in surviving the competitive market environment of this newly emerging industry, other cooperatives were soon able to keep

Fig. 3.2 Estimated share of raw material in Dutch paper production, 1870–1895 (*Source* Own calculation based on De Wit [1990, pp. 50–54])

[49] Dollard is a region in the federal state of Groningen.

[50] Bouwens (2004, p. 64).

[51] Ibid.

[52] Hopster (1992).

[53] Bouwens (2004, p. 65).

up with the speculative branch due to different reasons. For one, state-initiated land reclamation for agricultural usage and the rising pressures among farmers to utilize artificial fertilizers pushed the continuing over-production of straw.[54] Secondly, worldwide demand for straw-board kept growing, while the Dutch straw-board industry remained inimitable.[55] Lastly, the legal organization of these cooperatives secured the availability of raw materials, based on the company shares held by each farmer.[56] In turn, rising straw prices or shortages in raw materials had little to no impact on the cooperatives and their production output. Understandably, these cooperatives had very different market strategies and interests to those of the speculative companies, who aimed at lowering straw prices to reach higher profit margins. Thus, mutual cooperation between these two camps to strengthen the strawboard industry's international competitive advantages during crisis-ridden years was unthinkable.[57] Even though a few cooperative attempts had been made during the Dutch strawboard industry's 100-year history (1870–1970), these never achieved any viable success.

From 1880 onwards, wood fiber emerged as a feasible alternative to cloth and straw. Despite the obstacles Dutch paper makers encountered due to the scarcity of wood fiber domestically, it eventually emerged as the primary raw material for paper production in the Netherlands by 1895. This transformation was largely driven by financially robust paper companies like Van Gelder Zonen and KNP, who took the lead in exploring and adopting alternative raw materials to replace cloth and straw. In 1883, Van Gelder Zonen in Wormer successfully transitioned to using exclusively wood fiber, while KNP established their first paper mill based on cellulose in 1884.[58] The advancement of paper production through mechanization and the adoption of new raw materials were facilitated by networks that allowed the affluent gentlemanly capitalist class fraction to consolidate their financial resources and stay abreast of global advancements. Moreover, the collaborative efforts between them and the monarchs William I and II, as well as the post-1848 interconnected

[54] Ibid.

[55] Ibid.

[56] Ibid.

[57] Ibid., p. 67.

[58] De Wit (1990, pp. 48–50).

relationships with the liberal parliament, led to the (re)establishment of favorable competitive structures, at the detriment of smaller-scale, handmade paper producers in the Netherlands.

By the end of the nineteenth century, the Dutch paper industry experienced a surge in profitability due to advancements in straw-board production and the adoption of wood fiber by companies like KNP and VGZ. Similar effects of industrialization were observed not only in the paper industry but across various sectors in the Netherlands and abroad. The emergence of large-scale corporations had a detrimental impact on small and medium-sized enterprises throughout different industries. This decline was particularly evident in handmade paper production, initially in the Zaanstreek region and later in the Veluwe region. Figure 3.3 illustrates the decrease in the number of paper mills in both regions by 1903, alongside the thriving strawboard production in Groningen and the sustained paper production in Maastricht. By 1869, the three paper mills in Maastricht alone produced six times more white paper than all 21 paper mills combined in the Apeldoorn region.[59] These regional competitive dynamics in nineteenth-century Dutch paper production were heavily influenced by a fundamental shift in energy sources and industrial structure, characterized by a significant decline in the number of industrial windmills in the Zaanstreek region.[60]

The gentlemanly capitalist class fraction responded to the economic downturn in the Zaanstreek, re-distributional policies, and low protectionism by reinforcing their national paper cartels to ensure ownership and increase profits.[61] These cartel structures were a landmark of "[t]he specific structure of Dutch industry with its strong position of the family firms, the lack of raw materials and the absence of a heavy industry determin[ing] the way collusive practices appeared in order to limit risk and uncertainty for the entrepreneur".[62] Despite successfully exposing and condemning the cartel of handmade paper producers, KNP and VGZ themselves continued to rely on familial connections and networks to not only ensure the success and stability of their companies but also to pursue

[59] De Wit (1990, p. 58).

[60] Davids (2006, p. 550).

[61] Van Zanden and Van Riel (2004, p. 147).

[62] Dankers and Bouwens (2004, p. 13).

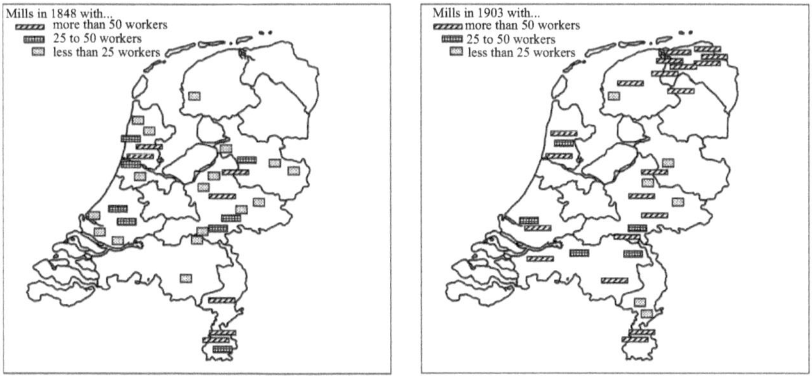

Fig. 3.3 Paper mills in the Netherlands in 1848 and in 1903 (*Source* Own illustration based on Everwijn [1912, pp. VI–VII])

their goal of centralizing the entire Dutch paper industry.[63] Though the legal structure of these family firms gradually transitioned from *rederijen* to non-listed limited companies, ownership, and managerial structures in the Dutch paper industry and broader Dutch economy continued to be organized through significant degrees of nepotism.[64] This nepotism manifested in the dominance of entrepreneurial dynasties by the end of the nineteenth century.[65] A prime example, the Dutch paper industry "was dominated by Van Gelder and Zon [VGZ], a 100 percent family owned business which produced nearly half of the total Dutch output, and KNP in Maastricht, [...] owned and managed by a group of Dutch and Belgian families".[66]

In 1904, the government collaborated with numerous smaller companies (around 39 paper mills) to establish the *Koninklijke Vereniging van Nederlandse Papier en Kartonfabrieken* (VNP—The Royal Association of Dutch Paper and Board Mills). This organization served as a lobby group, facilitating the exchange of information regarding raw material markets,

[63] Ibid., p. 11.

[64] Davids (2006, p, 562).

[65] Schrauwers (2010, pp 776–778).

[66] Dankers and Bouwens (2004, p. 15).

tariffs, and production capacity in other regions.[67] The establishment of the VNP aligned with the prevailing trend of corporate governmentality and was integrated into an intricate network of interconnected corporations.[68] This network extended the influence of the monarchy into civil society, the financial services sector, trade, and national industrial production. VGZ and KNP did not have any interest in joining the association. Instead, they had their own agreements in place to reduce the costs of paper production. Surprisingly, even without the participation of these two major industry players, the VNP managed to successfully establish lower price arrangements for raw materials, including straw, by exploiting peasants. While the smaller companies collaborated to exert control over raw material prices, the two dominant industry players, VGZ and KNP, continued to pursue their monopolistic market strategies.[69] Both these intra-industry and state-industry networks furthered the wealth of the gentlemanly capitalist class fraction by ensuring the perpetuation of labor exploitation, both in colonial and local contexts.

3.3 Modern Labor Exploitation

Comparably late, namely from 1820 onwards, the modern proletariat emerged in the Netherlands, as a result of the gradual annexation and privatization of the common land.[70] The annexation of common land, carried out by provincial generals acting on behalf of the monarch, primarily targeted rural areas in the hinterland of the Northern Netherlands (excluding Holland), which was originally leased by peasants from local nobles and landowners.[71] By stripping peasants of their rights to use the common land, the government was able to exert various forms of control.

Firstly, it facilitated the creation of a proletarian class, as peasants no longer owned the means of production and were compelled to sell

[67] Ibid.

[68] Schrauwers (2010, p. 754).

[69] Dankers and Bouwens (2004, p. 15).

[70] Federici (2004, p. 28).

[71] Van Zanden and Van Riel (2004, p. 130).

their labor in exchange for wages.[72] Secondly, it provided the government with a comprehensive understanding of the available landmarks for constructing infrastructure, which enabled the intensified exploitation of natural and labor resources. Thirdly, through land privatization, the government ensured the commodification and increased monetary value of the land, leading to the commercialization of agricultural production.[73] This process of dismantling and destroying the common land, driven by harsh capitalist practices, came at the cost of brutally suppressing the resistance of the *keuters* (common land peasants).[74]

Between 1860 and 1890, the number of industrial employees in the Netherlands doubled, reaching a total of 600,000 workers.[75] This significant growth in industrialization led to the emergence of new forms of labor-capital relations. Starting from 1870, waves of unionization occurred, along with more radical forms of workers' organization as active responses to the dominant powers at play.[76]State initiatives, nonetheless, continued to dismantle labor security and rationalize the wage system further in order to promote the investment interests of the gentlemanly capitalist class fraction. As a result, a marginalized and impoverished social class rose, the *lumpen proletariat*. With a growing fear of the dominant capitalist class fractions about the potential formation of uncontrollable workers' uprisings, similar to those witnessed in other countries during this time, the labor union prohibition law was finally abolished in 1869.[77] As a result, collective working agreements gradually spread across various sectors. Nonetheless, information regarding nationwide working conditions, wage scales, and employment opportunities remained limited and inaccessible to the majority of the proletariat.[78]

Workers faced daily exploitation under harsh labor conditions, enduring 12-hour workdays and experiencing sudden unemployment and unpaid wages during times of energy or raw material shortages. These

[72] Federici (2004, p. 29).

[73] Ibid.

[74] Van Zanden and Van Riel (2004, pp. 130–133).

[75] Dankers and Bouwens (2004, p. 16).

[76] Ibid., pp. 251–252.

[77] Van Zanden (2005, p. 70).

[78] Bourrinet (2016, p. 16) and Knippenberg and Pater (1990, p. 125).

challenging circumstances made it difficult for workers to organize themselves against the combined influence of the gentlemanly capitalist class fraction and state officials.[79] The state-led industrialization of agriculture and local industries further exacerbated the already precarious situation of peasants. A notable example of this was observed in the speculative practices of entrepreneurial factory owners and the pressure exerted by members of the VNP, which compelled peasants in the straw sector to sell their surplus straw at continuously declining prices.[80] In response, peasants formed cooperatives to increase the prices of their excess produce. However, mounting market pressures resulting from increased capital investments in the straw industry, driven by speculative companies, transformed these cooperatives into selling arrangements dominated by the highest-bidding middlemen (commissioners), ultimately leading to their bankruptcy.[81]

Another problem feeding into the resistance of workers in the Dutch paper industry and the Dutch economy more generally was that, from the formation of the Workers' League of Holland in 1871 onwards, Dutch unionism resembled the institutionalized fight for the *right to work*, instead of structural resistance to labor exploitation in industrialized capitalism. Therefore, independent anti-war, anti-work, and anti-colonial associations started forming around young collectives mostly based on the Marxist and anarchist idea(l)s, such as the Mokers group in 1904.[82] Herman J. Schuurman was an important figure in the rebellious youth wing of the Dutch libertarian movement, who edited, together with other comrades, *de Moker* magazine and organized nationwide meetings under the umbrella idea of abolishing work.

> This social system, capitalism, is based on the act of work; it formed a class of people, that have to work—and a class of people, that don't work. The workers are forced to work, because if not, they will have to starve. "Because," the owner teach us, "he who doesn't work, will not eat," and they claim that their calculating and gathering of profits is also work.[83]

[79] De Vries (1957, p. 331).

[80] Bouwens (2004, p. 168).

[81] Ibid., p. 170.

[82] Van Daele (2013).

[83] Schuurman (1924, *italics in original*).

Regrettably, the possible ties between anti-capitalist and anti-militarist workers' groups and laborers in the Dutch paper industry, as well as in other domestic industries, remain largely unrecorded owing to the considerable suppression by the state. The state employed measures aimed at marginalizing such organizations until their eventual defeat. Throughout time, both monarchs (William I and II) and the subsequent liberal parliament consistently opposed assisting workers' rights and their struggles. Workers' organizations, often targeted by networks of the dominant class fractions, such as the chamber of commerce, found their reasonable demands regularly rejected.[84] The legalization of unions in 1869 followed a similar strategic approach, serving as a preventive measure to avoid the emergence of more substantial workers' uprisings. Overall, the prevailing attitude was one of resistance toward supporting workers' rights and accommodating their demands.

To sum up, during Dutch monarchic liberalism the interests of powerful class fractions aligned in a way, which changed the characteristics of the Dutch paper industry drastically, away from peasant-based, handmade, small-scale paper manufacturers to the dominance of a large-scale, corporate paper industry (see Table 3.1). Subsequently, the networks within and between the dominant class fractions only enabled the continuity of national industries, not their catching up with the industrialization levels of neighboring countries. Thus, building Industria remained without success in terms of boosting the Dutch paper industry's international competitiveness.

[84] Davids (2006, p. 574).

Table 3.1 Comparing Dutch paper production regions, nineteenth century

	Zaanstreek/Maastricht	*Veluwe*	*Groningen*
Technology	– Steam-run paper machines – Production of 'endless' paper – Usage of wood fibers	– Continuing handmade paper production using cloth and based on traditional hamerbak – After 1890 either shut down or bought out	– Innovative use of straw as raw material – Board production only (no paper)
State-industry relations	– Network between gentlemanly capitalists and monarchs, later with liberal cabinet	– Little to no relations with state or local authorities	– Support in searching for innovative raw materials by state-organization (Beerta)
Competition and cooperation	– Stark concentration of industry – (Institutionalized) cartel structures	– Cartel-structures to maintain handmade Dutch paper tradition (Marten Orges)	– Stark competition between gentlemanly farmers and strawboard cooperatives
Labor-capital relations	– Rising proletarianization	– State-driven industrialization of agriculture	– Proletarian laborers work for gentlemanly farmers
	– Illegalized unionization – Formation of anti-war, anti-work, and anti-colonial workers' associations	– Switched to mainly proletarian laborers, similar to Zaanstreek region	– Self-employment in straw board cooperatives

References

Aglietta, M. (2000). *A theory of capitalist regulation: The US experience*. Verso.

Archieven, G. (2015). *Groninger Maatschappij van Landbouw*. Retrieved May 31, 2023, from https://www.archieven.nl/nl/zoeken?mivast=0&mizig=210&miadt=5&miaet=1&micode=405&minr=1449629&miview=inv2

Bourrinet, P. (2016). *The Dutch and German communist left (1900–68): 'Neither Lenin nor Trotsky nor Stalin!'—'All workers must think for themselves!'*. Brill.

Bouwens, B. (2004). *Op papier gesteld: De geschiedenis van de Nederlandse papier- en kartonindustrie in de twintigste eeuw*. Boom.

Brandon, P. (2011). Marxism and the 'Dutch Miracle': The Dutch Republic and the transition-debate. *Historical Materialism, 19*(3), 106–146.

Dankers, J., & Bouwens, B. (2004, May). *Behind the clouds: Cooperation in Dutch business around 1900*. Retrieved September 11, 2023, from https://bintproject.nl/textfiles/2004_bouwens-dankers-behind.pdf

Davids, K. (2006). The transformation of an old industrial district: Firms, family, and mutuality in the Zaanstreek between 1840 and 1920. *Enterprise and Society, 7*(3), 550–580.

De Groot, G. (2001). *Fabricage van verschillen: Mannenwerk, vrouwenwerk in de Nederlandse industrie (1850–1940)*. Aksant.

De Vries, B. W. (1957). *De Nederlandse papiernijverheid in de negentiende eeuw*. M. Nijhoff.

De Wit, O. (1990). *Papierfabricage in nederland in de negentiende eeuw van molen naar fabriek*. Stichting Historie der Techniek.

Everwijn, J. C. A. (1912). *Beschrijving van handel en nijverheid in Nederland*. Gebrs. Belinfante.

Federici, S. (2004). *Caliban and the Witch*. Autonomedia.

Hopster, J. (1992). *Strokarton: Een korte beschrijving van de opkomst en teloorgang van een specifiek Noord-Nederlands product*. Retrieved June 2, 2023, from http://demensenvandestrokarton.nl/strokarton-een-korte-beschrijving-van-de-opkomst-en-teloorgang-van-een-specifiek-noord-nederlands-product/

Hunter, D. (1978). *Papermaking: The history and technique of an ancient craft*. Courier Corporation.

Jonker, J. (1996). *Merchants, bankers, and middlemen: The Amsterdam money market during the first half of the 19th century*. NEHA.

Knippenberg, H. & Pater, B. C. (1990). *De eenwording van Nederland: Schaalvergroting en integratie sinds 1800*. Sun.

Martins, P. M. (2013). Left and right in the nineteenth century ideological debate: The case of Portugal (1820–1910). In J. C. Rosas & A. R. Ferreira (Eds.), *Left and right: The great dichotomy revisited* (pp. 140–164). Cambridge Scholars Publishing.

Milanovi, Z. (2013). *The role of the monarchy in the Netherlands*. Retrieved May 31, 2023, from https://www.marxist.com/role-of-monarchy-in-netherlands.htm

Pfann, G. A., & Van Kranenburg, H. (2003). Tax policy, location choices, and market structure. *The Journal of Law and Economics, 46*(1), 61–84.

Schrauwers, A. (2010). "Regenten" (gentlemanly) capitalism: Saint-Simonian technocracy and the emergence of the "Industrialist Great Club" in the mid-nineteenth century Netherlands. *Enterprise and Society, 11*(04), 753–783.

Schuurman, H. J. (1924). *For work is a crime*. De Moker.

Te Velde, H. (2008). The organization of liberty: Dutch liberalism as a case of the history of European constitutional liberalism. *European Journal of Political Theory, 7*(1), 65–79.

Van Daele, E. (2013). *De Moker group: The rebellious youth in the Dutch libertarian movement of the roaring twenties.* Uitgeverij de Dolle Hond.

Van Zanden, J. L., & Van Riel, A. (2004). *The strictures of inheritance: The Dutch economy in the nineteenth century.* Princeton University Press.

Van Zanden, J. L. (2005). *The Economic History of the Netherlands 1914-1995: A Small Open Economy in the 'Long' Twentieth Century.* London, UK: Routledge.

CHAPTER 4

Corporatization of Paper Manufacturing

Abstract This chapter explores the development of the Dutch paper industry in the phase of Fordism between 1914 and 1980. Following periods of economic downturn and uncertainty brought about by the First and Second World Wars, there was a concerted effort towards state-led restructuring and consolidation to rejuvenate national industries. During Fordism, cartel structures emerged as crucial forms of cooperation, while corporatization and mergers and acquisitions became tools for industry consolidation. The 1950s and 1960s witnessed a significant rise in unionism, but the economic downturns of the 1970s led to its rapid decline. Notably, in this era, waste paper emerged as a profitable substitute for virgin wood fiber in Dutch paper production.

Keywords Fordism · Cartels · Industry concentration and consolidation · Unionism · Waste paper

> The whole 'mode of regulation' was organized under the dominance of big monopolies, closely linked to the state, and allowed capitalism to expand in a relatively balanced, steady way.
>
> Thomas (2005, para. 15)

The Fordist accumulation regime constitutes the third phase of capitalism in the Netherlands from 1914 until 1980. Fordism is an accumulation regime with a distinct mode of regulation and set of technical

M. E. Ehrich, *The Dutch Paper Industry from 1580 to the Present*, Palgrave Studies in Economic History,
https://doi.org/10.1007/978-3-031-54324-1_4

and social relations of (re-)production, originating in the early twentieth-century US-American politico-economic context.[1] Fordism, as an ideal-type accumulation regime, involves a nationally based circuit of capital rooted in mass production and mass consumption.

> Fordism's virtuous circle involves rising productivity based on economies of scale in mass production, rising incomes linked to productivity, increased mass demand due to rising wages, increased profits based on full utilization of capacity, increased investment in improved mass production equipment and techniques, and a further rise in productivity.[2]

To understand the Fordist mode of regulation, one has to consider its institutional and organizational base, namely the involvement of state agents, class fractions, and labor-capital relations in the circuit of capital. First of all, the Fordist state can be characterized as a Keynesian-inspired welfare state, which secures capital-intensive investments in mass production and fosters R&D ventures while at the same time ensuring mass consumption through safeguarding a minimum prosperity of the lower and middle classes.[3] To do so, labor policies involve the recognition of unions for wage bargaining, minimum wage legislations, and the development of welfare programs to ensure high consumption levels for the unemployed. Exemplary Fordist organizations either base their rising scales of productivity on trusts and cartel structures or on horizontally integrated corporations. In Fordism, surplus profit is foremost based on returns of investment, as higher productivity lasts only "until the innovation(s) become standard practice".[4] The expansion of corporate enterprises thus relies on the availability of private capital and reinvestment of profits. In Fordism, consumer credits, secured by central banks, are fundamental to increase purchasing power of middle and lower classes and are in turn hedged through "state credit policies [...] aimed at aggregate demand and full employment".[5] Furthermore, the interdependence of mass production and mass consumption manifests through the rising

[1] Jessop and Sum (2006, p. 59).

[2] Ibid., pp. 59–60.

[3] Jessop and Sum (2006, p. 62).

[4] Ibid., p. 61.

[5] Ibid., p. 62.

importance of commercial capital, being ever-present in the form of mass advertising, mass retailing, mass credit, and mass media.

4.1 Interim War Period 1914–1945

The transition from the phase of Dutch monarchic liberalism to Fordism in the Netherlands was a gradual process, which followed the political and economic uncertainties of the interim war period from 1914 to 1945. World Wars I and II affected the Dutch economy heavily, including its paper industry. Phases of economic booms and downturns appeared at ever-shorter intervals from 1914 onwards, as reflected in the production output of the Dutch paper industry (see Fig. 4.1). On the one hand, Dutch paper production depended on imported raw materials, and was, thus, strongly affected by the unstable political climate between its European trading partners. On the other hand, the export possibilities for Dutch paper rose with the shutdown of production facilities in Germany, France, and the United Kingdom during World War I.[6] Even though the Dutch paper industry quadrupled its profits during this time, the first half of the 1920s was marked by a decreasing demand for paper and a thorough shortage of coal throughout Europe, which also took its toll on the industry. Before the economic crisis started in 1929, bringing prices for paper products to a record low, the Dutch paper industry experienced a short revival in the second half of the 1920s. Nevertheless, by the 1930s, the industry tried to countervail immense losses caused by the Great Depression by forming cartel agreements, such as negotiations on raw material prices.[7] The Dutch government also reacted to the volatilities and capitalist crises, banning the export of raw materials from 1930 onwards.[8] As a consequence, the Dutch paper industry did recover to some extent from the economic devastation of the late 1920s and 1930s.

With the beginning of World War II, the industry once more faced the industrial and political uncertainties of wartime. Under German occupation, the paper industry, among many other Dutch industries, was fully annexed, solely serving German demand for paper.[9] A shortage of

[6] Bouwens (2004, p. 73).

[7] Ibid., p. 80).

[8] Dankbaar and Velzing (2013, p. 6).

[9] Bouwens (2004, p. 85).

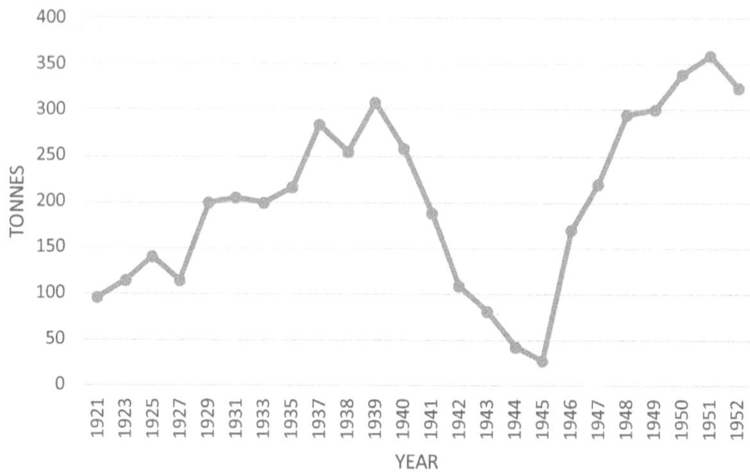

Fig. 4.1 Production of paper and board in tons in the Netherlands, 1921–1952 (*Source* Own illustration based on Bouwens [2004, pp. 77, 87, 115])

resources, including machinery, raw material, and fossil fuels, led to an immense decrease in Dutch paper production output and sites, and, thus, to a concentration of industrial activity. Many workers, who lost their jobs as a result of stand-stills or even shut-downs were recruited for the program of *Arbeitseinsatz*, a German program to forcefully relocate industrial experts from occupied territory to support German production.[10] Most of Dutch paper production either served the propaganda-program of Nazi-Germany or was heavily regulated to prohibit the production of, for example, specific newspapers, magazines, advertisements, and office materials. After a few months of war-activity within the occupied Dutch territory, the Dutch paper industry comprised only 50 companies, producing 350,000 tons of paper and board; by September 1944, the entire Dutch paper production had come to a standstill.[11]

[10] Ibid., p. 87).

[11] Ibid., p. 100.

4.2 Industry Consolidation

At the turn of the twentieth century, the importance of prior forms of network cooperation, namely mutual insurance contracts or *rederijen*, decreased as the majority of industrial companies turned to commercial insurance companies instead.[12] This does not mean though that network cooperation as a whole lost its importance for Dutch industries. On the contrary, previous forms of cartel structures were simply replaced by new forms of cartels, ultimately remaining the most important form of cooperation during the twentieth century in the Netherlands and abroad.

For example, the Nordic paper and pulp production, along with its trade, experienced intensive consolidation under the well-known SCAN-cartels, which were established in the 1930s.[13] These cartels posed a growing challenge for Dutch paper producers in terms of accessing raw materials and securing consumer markets. The increasing European integration through trade policies and the removal of tariffs also posed a threat to the international position of the Dutch paper industry by exposing it to greater international competition. Thus, also for the Dutch paper industry, the importance of network cooperation in the form of cartels and business interest associations became pivotal. A prime example, the VNP grew to flourish by the 1930s, fostering "[a]greements on prices, sales and production [to reduce] competition and provide[...] stability for the members".[14] These new forms of network cooperation and commercialized insurance slowly loosened the ties among members of the big Industria (either linked through family bonds or previous cartel structures) and established new ones with members of the growing finance and service sectors.[15]

This transition reflected a growing integration between industrial and financial interests within these networks. Starting after World War II, this transition reinvigorated European markets through redefining industrial and bank policies. From then on, the Ministry of Economic Affairs saw mergers "[...] as the most plausible way for improving efficiency, optimizing added value and stimulating effective corporate management

[12] Davids (2006, p. 569).

[13] Jensen-Eriksen (2017, pp. 4–5).

[14] Bouwens (2012, p. 195).

[15] Davids (2006, pp. 570–571).

in weak and vulnerable industries".[16] As a consequence, banks started issuing long-term loans to industrial agents for mergers and acquisitions (M&As) and this incommensurable trend progressed from vertical integration to growing numbers of takeovers. Before 1950, the main industrial region of the Netherlands, Zaanstreek, which had ceased in relevancy for the Dutch paper industry, was to a large extent closed for 'outside' investment capital.[17] Yet, during Fordism, the reliance of the industrialist class fraction on capital from within its (family) networks was soon replaced by finance capital from banks. This was partially due to the newly established 'old boys network', comprised of the supervisory directors of big corporations as well as banks, which dominated Dutch industries throughout the entire phase of Fordism.[18] These emerging networks represent early manifestations of the intersection between the industrialist and financial class fractions. Industrial agents began shifting their reliance away from internally accumulated capital and intra-industry network insurance. Instead, they increasingly relied on commercial insurance and the banking sector as a means of support.

Efforts of governmental restructuring to coordinate the productivity of the Dutch economy included the regulation of industrial usage of resources and price-fixing strategies, which made the 1950s a booming decade for the paper industry.[19] Accordingly, between 1950 and 1960 the Dutch paper industry produced up to 1.5 million tons of paper and board annually.[20] Soon machine numbers increased by 35 percent, comprising a total of 170 machines in use.[21] Until the 1960s, there was significant growth in employment rates and relatively few mergers and acquisitions in the industry.[22]

From 1960 onwards though, the state played a major role in supporting concentration in the case of the Dutch paper industry. Originally, the Dutch state had supported merger negotiations between the

[16] Bouwens and Dankers (2013, p. 1118).

[17] Davids (2006, p. 567).

[18] Horn and Vliegenthart (2010, pp. 63, 67).

[19] Bouwens (2012).

[20] Ibid., p. 129.

[21] Ibid., p. 131.

[22] Bouwens (2004, pp. 168–188).

two paper companies KNP and VGZ. These were stopped as both companies were simultaneously negotiating with the North American companies Crown Zellerbach and MacMillan Bloedel to secure joint investments into production facilities and plants.[23] Another factor, which influenced the stopping of merger negotiations between KNP and VGZ, was the fusion of Bührmann (originating in 1866 as a paper trading company) and Tetterode (a graphic trading company) in 1963. Following this mega-merger, Bührmann-Tetterode (BT) took over numerous other companies in various related sectors such as graphical machinery, stationary and envelopes, book-shops, publishing houses, and the toy-industry, paying the majority of these takeovers in cash, as was common in the 1960s when corporations still expanded through accumulated capital based on past profits.[24] Dutch businesses, overall, adopted strategies inspired by American and British examples, such as vertical integration and the formation of large corporations, to mitigate the high costs of declining international competitiveness due to high operating costs associated with labor, surplus production capacity, and stricter ecological standards compared to other European countries.[25]

As a response to the growing importance of international shared capital investments and mergers as well as the growing pressures from low-cost production in third-wave industrialized countries, the Dutch state was eager to restructure national sectors into concentrated industries.[26] Sector-restructuring strategies included the implementation of specific organizations under public law, which were independent advisory organizations, comprised of industrial experts. The best-known example of outsourcing state-led restructuring programs to industrial experts is the *Nederlandse Herstructureringsmaatschappij* (NEHEM, Dutch Industrial Reconstruction Corporation), founded in 1972 and designed after the British Industrial Reconstruction Corporation (IRC).[27]

During this time, BT as well as KNP "took advantage of [the] complex restructuring process of the national board industry and with the support

[23] Bouwens (2012, p. 198).

[24] Brouwer and Steijn (1976, pp. 62–63, 72).

[25] Bouwens (2003, p. 10; 2004, p. 188).

[26] Langdon (1981, p. 768).

[27] Langdon (1981, p. 768) and Van Zanden (2005, p. 141).

of the government acquired the lion's share of the solid board producers".[28] Gradually, joint ventures between KNP and BT started, leading to a growing concentration of the Dutch paper industry by 1970. To cope with the growing concentration, VGZ was forced into liquidation, which led to its bankruptcy after a request for financial support was denied by the Dutch state in 1981.[29] Interestingly, already in 1975, the government agency *Investerings- en Ontwikkelingsmaatschappij voor Noord-Nederland* (NOM, Investment and Development Company for the Northern Netherlands) had set up a joint venture with BT and KNP, named Kappa, a comparably big industrial paper plant for Dutch standards at that time. NOM held 49 percent of the shares of this new plant, which enabled Kappa to increase its production capacity by 60 percent.[30] A total of 115 million guilders of public money were invested into this state-industry joint venture.

With its efforts, the Dutch government actively served the interests of the industrialist and managerial class fractions in order to rehabilitate those particular industries, which focused on national demand, were internationally depreciated and outright shrinking. This is also the case for the Dutch paper industry, which had declined by 20 percent in its overall production capacity from 1967 to 1976.[31] Concurrently, the composition of transnational corporations in the Dutch paper industry changed from large, vertically integrated corporations like BT (NL), KNP (NL), MacMillan (CAN), and Crown Zellerbach (USA) in the 1960s, who mainly participated in joint ventures, to corporations following strategies of diversification and acquisition like Feldmühle (GER), CCA (USA), Reed (UK), BPB (UK), and Enso Gutzeit (FIN) in the 1970s.[32] As a result of increasing concentration and mounting pressure from the tin and plastics sectors, smaller companies within the Dutch paper industry gradually disappeared while transnational corporations (TNCs) emerged.[33]

[28] Bouwens (2012, p. 201).

[29] Ibid.

[30] Brouwer and Steijn (1976, p. 80).

[31] Bouwens (2004, p. 178).

[32] Bouwens (2003, p. 10).

[33] Ibid., p. 11.

During the 1970s, the Dutch paper industry encountered three primary obstacles. Firstly, it faced the critical need to find alternatives to expensive imported raw materials, specifically wood fiber. Secondly, due to the increasing speed and scale of the production process, overproduction occurred, necessitating the exploration of new markets. Lastly, the industry had to contend with rising energy and oil prices.[34] Despite these challenges, businessmen, consultants, bankers, and politicians assessed the industry's worth, agreeing that the Dutch paper industry had a right to existence.[35] Consequently, the Ministry for Economic Affairs invested heavily in the revitalization of the industry by encouraging industrial concentration in the form of M&As and by financially supporting the VNP, so the Dutch paper industry could rise to all three challenges.[36]

In addition, so called crisis cartels, which were tolerated at EU-level and state-co-orchestrated at the national level, became of crucial importance to European industries more generally, and the Dutch paper industry more specifically.[37] In the case of the Dutch paper industry, crisis cartels helped organize state-industry projects on energy cost reduction as well as on supporting the independence of the Dutch paper industry from raw material imports.[38] By means of crisis cartels and state subsidies, which were tolerated by the EU Commission as "rescu[ing] industrial sectors in despair", the waste paper had become the most important resource of Dutch paper production in 1970, by far replacing pulp usage.[39] This increase in the use of waste paper secured the import-independence of the Dutch paper industry alongside desirable improvements in terms of ecological pollution. Also as a means to maintain profitable prices, secure markets and restrict supply, cartel strategies continued to be of importance to the Dutch paper industry. In fact, rising EU-level pressures for implementing anti-cartel laws more forcefully within the Netherlands were downplayed or outright ignored by the cabinet and the Netherlands continued to be considered a "cartel

[34] Bouwens (2012, p. 198).

[35] Bouwens (2004, p. 204).

[36] Bouwens and Dankers (2012, p. 18).

[37] Buch-Hansen and Wigger (2011, p. 96).

[38] Bouwens (2004, pp. 132, 199).

[39] Buch-Hansen and Wigger (2011, p. 30), Jensen-Eriksen (2011, p. 200), and Van Veen-Groot et al. (2001, p. 31).

paradise" throughout the 1980s.[40] Next to the politically encouraged M&As, cartel practices continued to blossom, ultimately helping the Dutch paper industry to achieve its booming phase of the 1980s.[41]

Even before the installation and exemption of the 1970s crisis cartels, cartel agreements had characterized Dutch industrial activities for decades. Cartels exist as different forms of cooperation and are not always in opposition to competition.[42] Cooperation in the form of cartel agreements was a vital part of Dutch business culture before *and* after the official enactment of competition laws in 1956.[43] While certainly not among the most cartelized industries in the Netherlands between 1962 and 1980, the Dutch paper industry does nevertheless remain comparably steady in its numbers of cartel agreements compared to other industries, ranging between a maximum of about 35 in 1965 to a minimum of 18 in 1980.[44]

In the Netherlands, cartels frequently went unpunished due to the requirement of proving potential economic damages before initiating investigations into collusive practices.[45] This meant that the burden of proof rested with the accuser rather than the accused, indicating the presence of protectionist elements in Dutch competition law during that period. The Dutch paper industry participated in various forms of cartels and gentlemen's agreements in the Netherlands from 1962 to 1980. For instance, in 1962, the leading cartel form in the Dutch paper industry (as well as across most industries) was price-fixing cartels, amounting to 42 percent.[46] This form even rose by 5 percentage points until 1980. Followed by allocation cartels with 18 and condition cartels with 17 percent, both these cartel forms actually decreased by 7 and 4 percentage points respectively until 1980.[47] The least common cartel forms in 1962,

[40] Bouwens and Dankers (2014, p. 59).

[41] Sluyterman (2013, pp. 214–219).

[42] Fear (2006, pp. 1, 3).

[43] Tweede Kamer der Staten-Generaal. *Wet economische mededinging* (1965).

[44] Bouwens and Dankers (2010, p. 763).

[45] Siraa (2016, p. 77).

[46] Bouwens and Dankers (2010, p. 764).

[47] Ibid.

quotas (3%) as well as rebate and exclusive trade (3%) nearly doubled until 1980.[48]

These developments align with the decline in the number of Dutch paper producers and an overall drop in production figures for the first time since World War II.[49] During this period, the Dutch paper industry faced high production costs and escalating international competition. As the perception of the economic crisis shifted toward a crisis of the embedded liberalist compromise, supranational anti-cartel legislation became more stringent.[50] Approximately 40 percent of all cartel prosecution procedures conducted by the European Commission between 1970 and 1990 were related to Dutch markets.[51] However, the mentioned statistics reveal only a slight reduction in the proportion of various forms of cartels in the Dutch paper industry from 1962 to 1980. This confirms that the achievements of anti-cartel legislation only resulted in sporadic successes.

4.3 Waste Paper

The 'discovery' of waste paper as a viable, profitable substitute for virgin wood fiber depicted a general trend in European paper production from the 1950s onwards.[52] In the specific context of the Dutch paper industry, it led to new industrial policies, the restructuring of waste management, and the adaption of innovative technology. Industrial policies ranged from financial subsidies for using waste paper in the cardboard and paper production to financial subsidies for warehouses. This latter state policy was called the EXPOVA agreement, which allowed paper producers to store waste paper in times of excess, and to use or sell it in times of shortage.[53] The Dutch paper industry and the industrial policies of that era placed a significant emphasis on ensuring a constant and accessible supply of this new and essential resource. To achieve this goal, the *Rijksbureau* (National Bureau) actively encouraged the public collection of

[48] Ibid.

[49] Bouwens (2012, p. 199).

[50] Buch-Hansen and Wigger (2010, p. 32).

[51] De Jong (1990, p. 215).

[52] Bouwens (2004, p. 132).

[53] McKinney (1994, p. 14).

waste paper through various initiatives. One such program was the *Jeugd Actie Papier Inzameling* (JAPI, Youth Action Paper Collection), which encouraged children to collect waste paper by offering incentives such as toys or recycling subsidies for voluntary organizations and local governments. These efforts aimed to incentivize and facilitate the collection of waste paper, ensuring its availability for the Dutch paper industry.[54] From the 1950s onwards, the use of waste paper in Dutch paper and board production rose substantially. By 1992, the paper producers in the Netherlands utilized up to 70 percent of possible waste paper and recovered around 55 percent of all waste paper.[55]

The continuous efforts to optimize waste paper management through industrial policies and the subsequent stimuli for paper consumers were accompanied by the adaption of suitable technology. In order to achieve high levels of waste paper usability, the production of cardboard rather than high-quality paper became essential to the Dutch paper industry.[56] Accordingly, waste paper preparation systems as well as parts of the paper machinery were adapted to the challenges posed by this new raw material. While "a wastepaper line of the 50s [...] would have the basic components of the 'state of the art' plant today", recovered fibers did indeed demand a different treatment during the paper and board production processes.[57] As an extension to the original Fourdrinier design, a section was incorporated into the paper machine to process recovered paper. In this stage, the recovered paper is mixed with water and chemicals and subjected to heat to transform it into loose fibers. In a subsequent step, the pulp mixture undergoes a screening process to eliminate undesirable substances like plastic, glass, ink, and sand.[58] To this day, pulp from waste paper remains less pure than virgin wood fiber pulp despite all efforts to further innovate the recovery process. Nevertheless, the introduction of waste paper in the Dutch paper production cycle not only relaxed its dependence on importing virgin wood fibers and raw material logs but also served as a viable investment outlet. These technological innovations certainly yielded profit and growth for the capital-owners of the Dutch

[54] Bouwens (2004, p. 112).

[55] McKinney (1994, p. 11).

[56] Bouwens (2004, p. 133).

[57] McKinney (1994, pp. 49–50).

[58] Marsidi (2008, p. 25).

paper industry, but they were not necessarily improving the situation of its laborers.

4.4 The Rise and Fall of Unions

After the end of World War II, many European states started regulating market forces and implementing wide-ranging social welfare programs as an effective strategy to guarantee rising productivity and full employment.[59] The phase of Fordism in the Netherlands is characterized by a state, which propels industrial development through creating large-scale national industrial champions. In 1948, the Central Bank of the Netherlands was nationalized and the national organization of employers and employees, called *The Stichting van de Arbeid*, was involved in the implementation of the guided wage policy.[60]

Trade unionism in the Netherlands achieved its first nationwide success in 1914 with a collective agreement on the eight-hour working day, the illegalization of child labor, and different regulations on working conditions.[61] From now on, most sectors, including the Dutch paper industry, switched from a two-shift to a three-shift system, partially releasing workers of their harsh working conditions. Also, the trade unions' wage bargaining of the twentieth century originates in the early success of workers' organizations. "The late industrialization and the parallel movement of pillarization produced a rather complex structure of the trade unions, characterized by strong national federations and, at times, fierce competition between the socialist and confessional trade unions".[62] This led to a weakening of the communist unions, in comparison to the right-wing, conservative and socialist unions, which remained strongly represented in the pillarization. Even though unions were partially re-appropriated by the corporate economy from 1919 onwards and taking seats in the High Council of Labor, maintaining ever-closer ties with affiliated political parties, they continued to pose a liable threat to the dominant class fractions. Hence, new forms of employers' network cooperation developed, directly targeting workers' liberation movements.

[59] Van Zanden (2005, p. 128).

[60] Ibid., p. 129.

[61] Van Damme and Peters (1994, p. 19) and Van Leeuwen (1997, p. 767).

[62] Van Zanden (2005, p. 70).

A prime example, the chamber of commerce—the formal network cooperation of the industrialist class fraction at that time—was renewed into the *Zaansche Werkgevers Vereeniging* (Zaanstreek employer association). Herewith, employers tried to "combat the workers more effectively, and especially [...] prevent firms in different branches of industry from being played off against each other".[63] To do so they copied the organizational structure of the workers' unions, spanning a multitude of sectors and regions. In 1919, the year the Zaanstreek employer association was founded, it counted 14 companies from nine different branches; in 1920, its membership numbers had increased to 78, and in 1922 to 100, by this time spanning the entire Netherlands.[64] One result of the growing dominance of the Zaanstreek employer association was, for example, that unemployment insurance became state-controlled comparably late, namely after the German occupation in 1943.[65] Another direct result of the increase in the dominance of the employers' network organization was that strikes decreased substantially from the 1930s onwards. In addition to joining the Zaanstreek employer association, employers of the Dutch paper industry also associated against the state and workers by joining the already 1904-founded lobbying organ VNP.[66] The strong organization of employers in the first half of the nineteenth century throughout the majority of sectors in the Netherlands peaked in the exclusion of the communist-led *Eenheidsvakcentrale* (the umbrella organization of the communist trade unions) from the centralized wage negotiations, which concomitantly ended the most bitter conflicts in the cotton industry that had criticized the re-appropriation of unionism by the corporate system and the unions' inability to truly represent the proletariat.[67]

From the 1950s onwards, the rise of Fordism in Western industrialized economies generally and in the Netherlands more specifically, demarcated a change in labor-capital relations during Fordism. Fordism describes the "parallel restructuring of both the technological and organizational basis of the production process and the lifestyle of the wage earners".[68]

[63] Davids (2006, p. 575).

[64] Ibid.

[65] Ebbinghaus and Visser (1999, p. 139).

[66] Van Damme and Peters (1994, p. 20).

[67] Van Zanden (2005, p. 78).

[68] Koch (2004, p. 15).

Rooted in early slavery and exploitation practices, Taylorist mass production of organizing working procedures as well as controlling the workers alongside the assembly line are pivotal strategies in Fordism. In Fordist times, profits were highly dependent on consumer demands; thus, high and growing wages are essential to write off fixed capital quickly.[69] In effect, workers were able to afford what they produced for the first time in capitalist history.[70] Additionally, relative wage costs rose strongly due to, among other factors, the introduction of the five-day workweek.[71] Exemplary of the rising consumption levels at that time, also Dutch paper consumption increased from one million tons to three million tons between 1961 and 1990.[72] Especially a steep rise in everyday products such as toilet paper and napkins can be noted for that period of rising purchasing power of the working class.[73]

The state played an important role in securing the Fordist regime of accumulation, "[...] foster[ing] growth and productivity agreements between employers' organizations and trade unions by promoting capital accumulation through public infrastructure spending and permissive credit and monetary policies".[74] Accordingly, the interests of the socialists and conservatives, their corresponding unions, and the industrialist class fraction aligned well during that time. Embedded in the Fordist accumulation regime and industry-wide restructuring processes, unionism rose again. Even though "wage bargaining remained a highly centralized process", collective workers' agreements (CAOs) continued spreading during the 1950s as trade unions were strengthening their positions through mutual cooperation.[75] Consequently, the Dutch paper industry saw the inaugural signing of a CAO in 1950.[76] However, a significant portion of workers believed that their interests were not adequately represented in the collaboration between workers' unions and the industrial class fraction. As a result, they organized what are known as "wildcat

69 Ibid., p. 16.

70 Göttfert (2012, p. 102).

71 Van Zanden (2005, p. 133).

72 Van Veen-Groot et al. (2001, p. 30).

73 Van Damme and Peters (1994, p. 23).

74 Koch (2004, p. 16).

75 Van Zanden (2005, p. 76).

76 Van Damme and Peters (1994, p. 19).

strikes", which ultimately did not bring about the desired structural changes.[77]

Throughout the 1960s, rising labor productivity came to determine wages, which soon ceased in excessively high wage levels. First, these developments seemed to support the workers' consuming behavior and, thus, helped maintain the Fordist accumulation regime. Soon, the crises of the 1970s in combination with the comparably high wage levels in the Netherlands, led to a rapid decline in employment rates and, hence, lower labor productivity and union membership. Concurrently, with the decline of heavy industries, the service industry grew and the demand for two kinds of laborers increased: Highly educated ones, demanded in government employment, business services, and health care, as well as unskilled workers, demanded in sectors such as leisure and catering.[78] These developments mark the beginning of a changing labor market structure in the Netherlands, which is dominated by a stark rise in higher education personnel in the successive accumulation regime of post-Fordism.

4.5 When Endless Growth Still Seemed Possible

After a phase of economic downturn and insecurities during the First and Second World Wars, the historicization shows that state support, which varied greatly over time, is reformulated in the phase of Fordism, heralding state-led restructuring and concentration as solutions for reinvigorating national industries. Networks between industry and state took the forms of active state support for mergers and acquisitions, which quickly led to the almost complete internationalization of the Dutch paper industry in terms of ownership. Later in the 1970s, the state even actively supported the formation of crisis cartels to secure the profitability of national industries.

Network cooperation within the Dutch paper industry developed accordingly, making cartel structures one of the most important forms of cooperation within the industry. Through such close cooperation, the no-longer Dutch-owned paper mills were able to develop and implement the usage of waste paper as raw material. Herewith, their international competitiveness rose as resource dependence from Scandinavian and other

[77] Van Zanden (2005, p. 78).

[78] Ibid., p. 82.

large-scale pulp and paper manufacturing countries decreased. The search for alternative fibers and other technological innovations was thoroughly promoted by the state through financial aid and organizational support. With the introduction of Fordist working structures and strengthening of cooperation among the industrialist class fraction, illustrated for example in the rise of the chamber of commerce, labor unions gradually weakened.

Overall, the so-called 'golden years' of capitalism (1950 until the oil crisis in 1973) were not a single homogenous period of growth but rather marked by different shifts in the macro-economic development of the Netherlands as well as changing dominance of class fractions' interests. In fact, the developments in the Netherlands during Fordism depict a continuous discontinuity. The industrialist class fraction and political authorities overlapped in their interest to reinvigorate Dutch industrial sectors after World War II. Thus, during the 1950s, state-led sector restructuring yielded substantial profits. These rising profits and the rapid industrialization of the 1950s in turn yielded overproduction in the 1960s.[79] At this time, the state shifted its strategies toward corporatization as M&As were seen as viable routes to cut production costs and deal with overproduction (see Table 4.1).

Consequently, economies of scale, rising wages, and increased international competition did not lead to a general cool-down of industries, but to ever-more capital investment, which in turn made exports grow substantially.[80] Only when wages began to outgrow productivity and inflation accelerated at the end of the 1960s, the economy grew to be "overheated" in the 1970s.[81] Scholars agree to some extent that Fordist regulation gradually erodes from this point onwards. Reasons for this are "various economic and social crises, increasing competitions on the international and global level, far-reaching technological innovations and sometimes drastic revisions in economic and social policies".[82] As will be shown in the following chapter, the Dutch paper industry underwent post-Fordist restructuring processes accordingly. Marked by financialization, flexibilization, and internationalization, the Post-Fordist

[79] Van Zanden (2005, p. 136).

[80] Ibid.

[81] Ibid.

[82] Koch (2004, p. 16).

Table 4.1 Comparing Dutch paper production regions, twentieth century

	Zaanstreek/Maastricht/Veluwe
Technology	– Industrialized production – Waste paper becomes pivotal resource
State-industry relations	– Substantial state subsidies, also in the form of joint ventures – State-led sector restructurings and corporatization efforts – Little to no prosecution of cartel practices – State support of consolidation strategies and M&As
Competition and cooperation	– Internationalization and further concentration of the Dutch paper industry – Ceasing importance of manufacturing industries for the Dutch economy – Cartel paradise
Labor-capital relations	– Strong employers' associations (e.g., chamber of commerce) – Weakening of (communist) workers' organizations through pillarization – The rise of Taylorist working procedures

accumulation regime promotes a supposedly new form of cooperation: Networks.

References

Bouwens, B. (2003). *The internationalization of the Dutch paper and board industry.* Unpublished paper presented at the BHC/EBHA Meeting, Lowell, Massachusetts, USA.

Bouwens, B. (2004). *Op papier gesteld: De geschiedenis van de Nederlandse papier- en kartonindustrie in de twintigste eeuw.* Boom.

Bouwens, B. (2012). The paper and board industry in the Netherlands, 1800–2000. In J.-A. Lamberg, J. Ojala, M. Peltoniemi, & T. Särkkä (Eds.), *The evolution of global paper industry 1800–2050* (pp. 191–209). Springer.

Bouwens, B., & Dankers, J. (2010). The invisible handshake: Cartelization in the Netherlands, 1930–2000. *Business History Review, 84*(4), 751–771.

Bouwens, B., & Dankers, J. (2012). *Tussen concurrentie en concentratie: Belangenorganisaties, kartels, fusies en overnames.* Boom.

Bouwens, B., & Dankers, J. (2013). Competition and coordination: Reconsidering economic cooperation in Dutch business, 1900–2000. *Revue Économique, 64*(6), 1105–1124.

Bouwens, B., & Dankers, J. (2014). From paradise to hell. Post-war cartel policy and practice in the Netherlands. ESKA: *Entreprises et Histoire, 3*(76), 58–71.

Buch-Hansen, H., & Wigger, A. (2010). Revisiting 50 years of market-making: The neoliberal transformation of European competition policy. *Review of International Political Economy, 17*(1), 20–44.

Buch-Hansen, H., & Wigger, A. (2011). *The politics of European competition regulation: A critical political economy perspective*. Routledge.

Commission of the European Communities. (1976). *A study of the evolution of concentration in the Dutch paper products industry*. http://aei.pitt.edu/41058/1/A5151.pdf

Dankbaar, B., & Velzing, E.-J. (2013). *Industriebeleid in Nederland sinds 1850: Industrialisatie, de -industrialisatie, reïndustrialisatie. Samen Toekomst Maken Met Techniek*. Retrieved September 11, 2023, from https://www.researchgate.net/publication/309011396_Industriebeleid_in_Nederland_-_Industrialisatie_de-industrialisatie_reindustrialisatie

Davids, K. (2006). The transformation of an old industrial district: Firms, family, and mutuality in the Zaanstreek between 1840 and 1920. *Enterprise and Society, 7*(3), 550–580.

De Jong, H. W. (1990). Nederland: Het kartelparadijs van Europa? *Economisch-Statistische Berichten, 75*, 244–248. https://esb.nu/nederland-het-kartelparadijs-van-europa/

Ebbinghaus, B., & Visser, J. (1999). When institutions matter: Union growth and decline in Western Europe, 1950–1995. *European Sociological Review, 15*(2), 135–158.

Fear, J. R. (2006). *Cartels and competition: Neither markets nor hierarchies*. Harvard Business School.

Göttfert, C. (2012). *Gesellschaftliche Naturverhältnisse und hegemoniale Tendenzen der biobasierten Ökonomie am Beispiel niederländischer Politiken* (Diplomarbeit). Retrieved September 11, 2023, from https://utheses.univie.ac.at/detail/16679

Horn, L. C., & Vliegenthart, A. (2010). Corporate governance in Nederland - hoe bedrijven handelswaar werden. In J. Blom (Ed.), *De Kredietcrisis - Een politiek-economisch perspectief* (pp. 59–78). Amsterdam University Press.

Jensen-Eriksen, N. (2011). Industrial diplomacy and economic integration: The origins of all-European paper cartels, 1959–1972. *Journal of Contemporary History, 46*(1), 179–202.

Jensen-Eriksen, N. (2017). Creating clubs and giants: How competition policies influenced the strategy and structure of Nordic pulp and paper industry, 1970–2000. *Business History, 62*(5), 763–781.

Jessop, B., & Sum, N.-L. (2006). *Beyond the regulation approach: Putting capitalist economies in their place*. Edward Elgar Publishing.

Koch, M. (2004). European reactions to global challenges: The cases of the Netherlands, the UK, Sweden, Spain, and Germany. *Journal of International & Comparative Social Welfare, 20*(1), 15–22.

Langdon, S. W. (1981). North/South, West and East: Industrial restructuring in the world economy. *International Journal, 36*(4), 766–792.

Leeuwen, M. H. D. (1997). Trade unions and the provision of welfare in the Netherlands, 1910–1960. *Economic History Review, 50*(4), 764–791.

Marsidi, M. (2008). *Climate change and resource efficiency in paper production*. Utrecht University.

McKinney, R. (1994). *Technology of paper recycling*. Springer Science & Business Media.

Siraa, T. (2016). *Economische Zaken: Een ministerie in historisch perspectief*. Elsevier.

Sluyterman, K. (2013). *Dutch enterprise in the 20th century: Business strategies in small open country*. Routledge.

Thomas, M. (2005). *"Post-Fordism": Collapsing into the present*. Retrieved June 27, 2023, from http://www.workersliberty.org/story/2005/10/14/post-fordism-collapsing-present

Tweede Kamer der Staten-Generaal. *Wet economische mededinging* (1965). Tweede Kamer.

Van Damme, M., & Peters, H. (1994). *Projectbureau Industrieel Erfgoed Rapportenreeks: Papierindustrie (Report No. 11)*. http://www.industrieel-erfgoed.nl/publicaties/pie-rapporten

Van Veen-Groot, D. B., Nijkamp, P., & Van Den Bergh, J. C. J. M. (2001). A scenario study of globalization impacts on international transport and the environment: An application to the Dutch paper industry. *Journal of Environmental Planning and Management, 44*(1), 21–40.

Van Zanden, J. L. (2005). *The economic history of the Netherlands 1914–1995: A small open economy in the 'long' twentieth century*. Routledge.

CHAPTER 5

Transnational Capital and Paper Production

Abstract This chapter examines the progression of the Dutch paper industry during the post-Fordist phase, starting from 1980 onward. By highlighting both active and passive government interventions and economic restructuring within the context of the Dutch paper industry, the chapter challenges the notion that the state withdraws from regulatory involvement in the post-Fordist accumulation regime. State strategies, including the deregulation of market structures to facilitate widespread transnational ownership, neoliberal re-regulative initiatives involving closer collaboration between the public and private sectors for research and development, privatization of public sectors, active and passive financial support for industry demands, and the flexibilization of labor, all aimed at enhancing the international competitiveness of Dutch industries. The chapter clarifies the tensions that arise from the convergence of conflicting class fraction interests, with a focus on industry sustainability, emerging forms of labor precarity, and heightened competition.

Keywords Post-Fordism · De- and re-regulation · Transnational corporations · Deindustrialization · Financialization · Labor flexibilization · Circular economy · Managerialism · Hyper competition

M. E. Ehrich, *The Dutch Paper Industry from 1580 to the Present*,
Palgrave Studies in Economic History,
https://doi.org/10.1007/978-3-031-54324-1_5

> Post-Fordism may be summarised as the obverse of Fordism. [...] The old oligopolistic firms occupy a smaller proportion of the value chain and establish quasi-market, network relations with a myriad of suppliers, distributors and even direct competitors.
>
> Friedman (2000, p. 60)

> What you see now is, there are only a few family-owned companies, the rest are all in the hands of multinationals or private equity. In the last years it is more and more private equity, especially in the North of the Netherlands. People who spend money—risk capital investors—is where all the shit in the world comes from... and all these companies go bankrupt. What they do is, they buy you and after four, five years they sell you again and they cash millions, and millions and hundreds of millions off of it and don't care about the continuity of the mill. So you're better off to not fall into the hands of private equity.
>
> Manager in a Dutch paper mill

Post-Fordism is an accumulation regime with a distinct mode of regulation and set of technical and social relations of production, originating in the political-economic conjunctures of the 1970s.[1] The post-Fordist accumulation regime manifests itself in the Netherlands from 1980 onwards.[2] The transition from the Fordist to the post-Fordist accumulation regime is not linear and coherent, but rather irregular, contextual, and diverting across countries. Nevertheless, conceptual definitions of the main post-Fordist characteristics are essential for understanding the changes the Dutch paper industry underwent since 1980.

The main characteristics of post-Fordism are labor market flexibilization, deregulation and neoliberal re-regulation, transnationalization of production and global value chains, the rise of finance-led accumulation patterns, and deindustrialization.[3] During Fordism, many industries still focused on standardized mass production, which resulted in saturated consumer markets in the 1970s in first-wave industrialized countries.[4] Under post-Fordism, this focus changed in accordance with the rise of

[1] Jessop (1993, p. 8).

[2] Koch (2004, p. 16).

[3] Jessop (1994), Jessop and Sum (2006), and Overbeek et al. (2007).

[4] Jessop (1994, p. 258).

cheaper production capacities in newly emerging markets, such as China.[5] Product specialization and the flexibilization of commodity production, thus, became essential for manufacturing industries to compete in an increasingly globalized market. Consequently, post-Fordist manufacturing operations are based on flexible machines and market-driven, specialized niche products to ensure profitability.[6]

Also labor is gradually being flexibilized in post-Fordism. Rooted in national and supranational labor policies, labor is subordinated to market competitive forces and labor markets are reoriented toward the supply-side.[7] Labor policies foster "'flexi-waged, flexi-time, hire-and-fire, and outsourced jobs through self-employed or subcontracted labor, multi-skilling of core workers enjoying job rotation, job enrichment and team-work", internationalization of managers' and technocrats' career-paths, rising dominance of temporary labor-contracts, the individualization and erosion of working-class communities, and the weakening of organized labor; all of which are part of a wider program of deregulation and neoliberal re-regulation in the post-Fordist accumulation regime.[8] The deregulation of the public sector involves its privatization through selling state shares and listing state-run industries, such as the railway industry, the health sector, the energy sector, and telecommunication sector, on stock markets. Due to these developments, the state is assumed to be retrenching during post-Fordism, leaving previously public as well as private sectors to compete under free market principles without state intervention. At the same time, post-Fordism is marked by waves of neoliberal re-regulation, geared toward the promotion of the competition regime.[9]

Especially the deregulation of financial systems is guided by and itself intensifies finance-led accumulation.[10] Through its deregulation "finance became once more detached from the real economy and liquid capital

[5] Ibid.

[6] Jessop (2002, p. 98).

[7] Ibid., p. 210.

[8] Jessop (2002, pp. 98, 109).

[9] Buch-Hansen and Wigger (2010).

[10] E.g. Krippner (2005).

gained a new transnational mobility and hence exit power".[11] Productive capital situated within traditional manufacturing industries becomes dependent upon finance capital (mostly banks and offshore capital) to sustain its production processes. Financialization, thus, describes "a pattern of accumulation in which profits accrue primarily through [...] the provision (and transfer) of liquid capital in expectation of future interest, dividends, or capital gains" instead of previously dominant forms of accumulation, namely trade and commodity production.[12]

Concomitantly, finance and productive capital are transnationalized during post-Fordism.[13] The loosening of barriers for capital movement and the spatial decentralization of production ownership gives rise to transnational corporations (TNCs). TNCs thrive under the minimization of tax liabilities and the removal of trade and investment barriers, as the resultant transnationalization of production processes allows for the maximization of their global profits.[14] The dynamic interplay between large European and US-American TNCs on the global (financial) market excels competitive forces and, in return, yields the continuation of deregulation of cross-border transactions in the interest of capital agents.[15] Since the 1990s TNCs have outdated former multinational corporations (MNCs), as the majority of large-scale corporations not only operate across national regimes but are embedded in the societal, political, and economic frameworks of the countries they operate in.[16] Unlike other major entities like governments and labor groups, TNCs occupy a unique position: They have the capability to amass resources that allow them to actively participate in political activities while also possessing the ability to navigate multiple, distinct governmental policies, often exploiting regulations that best serve their interests.[17] Furthermore, TNCs' operations and structures changed under rising globalization in post-Fordism as previously

[11] Van Apeldoorn and Horn (2007, p. 83).

[12] Krippner (2005, pp. 174–175).

[13] Bratsis (2014, p. 116).

[14] Overbeek (1993, p. 259) and Wigger (2015, p. 122).

[15] Buch-Hansen and Wigger (2010, p. 33).

[16] Bélanger and Edwards (2006, p. 29).

[17] Ibid.

still centralized functions are now internationally dispersed, such as R&D and design.[18]

The shift in global profitability from productive to financial capital and the growing reliance of industrial manufacturing companies on liquid capital is essential for the ongoing process of deindustrialization, which marks so-called 'advanced economies' in post-Fordism. The rising dominance of specialization, continuous innovation, economies of scope, and rapid responsiveness to consumer markets accelerates the dominance of TNCs and the rise of the service sector, while manufacturing industries are relocated to China and other newly industrialized countries.[19] The process of deindustrialization becomes evident in first-wave industrialized countries through two key indicators: A substantial decrease in the contribution of manufacturing to the overall Gross Domestic Product (GDP) and a decline in the proportion of manufacturing jobs within the total employment sector.[20] Under the impact of deindustrialization, several consequences emerge and contribute to its intensification: A reduction in union membership, a decline in the political influence of the working class, the emergence of a more vulnerable and precarious workforce, and the erosion of security within the labor market.[21] Ultimately, the interplay between deindustrialization and national industrial policies focused on labor market flexibilization limits the ability of labor to effectively organize against (transnational) capital in the post-Fordist era.

5.1 Industrial Policy in Post-Fordism

A variety of scholars argue that the transition to post-Fordism is accompanied by a gradual disembeddedness of markets from states.[22] According to them, the past decades are demarcated by laissez-faire state strategies, hereby allowing finance capital to control national market developments as well as the global economy. This state retrenchment thesis follows the

[18] Lundan and Mirza (2011, p. 31).

[19] Jessop (1994, pp. 252ff.) and Jessop (2002, p. 99).

[20] Tregenna (2011, p. 5).

[21] Varga (2013, p. 443).

[22] Altvater (2009), Altvater and Mahnkopf (1997), Helleiner (1996), Hutton (2008), and Lacher (1999).

"key tenets of neoliberal ideology" by concluding that recent politico-economic developments, such as flexibilization, deindustrialization, and financialization are the foremost results of a 'silent' state.[23] However, the historicization of the Dutch paper industry reveals a different dynamic. Despite the transition to post-Fordism, state involvement and industrial policies persist, albeit in altered forms and with different objectives. These factors have played a crucial role in sustaining the Dutch paper industry. By studying the evolving regulatory approach since 1980 in the Netherlands, both direct and indirect market engagements of state authorities become evident.

During the 1980s, the Dutch paper industry experienced a notable expansion in terms of profitability, production output, and investments. Specifically, the number of high-capacity production machinery tripled within a period of eleven years, allowing for an investment of 12 percent working capital on average for meeting operational expenses and short-term debt obligations.[24] Three main factors contributed to the industry's growth at that time: New strategies of capital investment, reduced costs for raw materials (including energy), and decreasing labor costs. I will discuss each one of them at length with a focus on the involvement of the state.

First, already during the 1960s and 1970s the Dutch government initiated programs to foster foreign direct investment (FDI) in the national paper industry.[25] While comprehensive data on FDI in the Dutch paper industry during that specific period are not available, other discernible indicators shed light on the industry's development. These indicators include a noticeable rise in the number of mergers and acquisitions, the prevalence of new investment strategies, and expanding profit margins. These trends suggest that the industrial policy implemented during that time was primarily geared at industrial growth.

A compelling illustration of this can be seen in the efforts of the Ministry of Economic Affairs, which actively facilitated inquiries from US-based companies regarding potential plant locations and labor supply.[26] Additionally, the ministry provided various forms of support, such as

[23] Panitch and Konings (2009, p. 68).

[24] Bouwens (2012, p. 202).

[25] Bouwens (2003, pp. 7–8).

[26] Ibid.

investment subsidies, grants, accelerated depreciation schemes, and fiscal stimuli. These initiatives further bolstered the industry's development and underscored the efficacy of the industrial policy in place during that period.[27] With the rise in FDI in the 1980s, foreign investors gradually implemented new forms of industrial investment strategies, such as expansion, diversification, and the elimination of competitors, allowing corporations to expand their already strong market positions.[28] By the 1990s, strategies of horizontal concentration as well as forward and backward integration became dominant tools to reach even more low-cost market bases.[29] Hence, profit margins rose steadily and vertical integration, such as the control of suppliers, distributors, or wholesalers, secured viable market outlets for Dutch paper producers.[30]

During this time of corporatization and consolidation, foreign investors displayed a primary interest in expanding the production output of Dutch paper mills by capitalizing on economies of scale. The emphasis was on achieving higher efficiency and cost-effectiveness through increased production volume. As the industry progressed, Dutch paper mills began exploring avenues beyond mass production. They ventured into strategies that allowed them to distinguish their products from competitors and cater to specific market segments with unique needs. Thus, mass production eventually evolved into strategies that incorporated product differentiation and specialization in niche markets.[31] This shift expanded the industry's scope of operations and opened up new opportunities for growth and profitability.[32]

These trends in investment strategies were closely intertwined with the evolving politico-economic landscape of European integration during that period. In 1987, the Single European Act came into effect, eliminating investment and production barriers among member states while also aligning their respective national regulations to foster competition.[33] During the same period in the Netherlands, the center right Lubbers I

[27] Bouwens (2003, p. 9).

[28] Ibid., p. 11.

[29] Ibid., p. 19.

[30] Ibid., p. 13.

[31] Ibid.

[32] Ibid., pp. 13–20.

[33] Watkins (2005).

cabinet (1982–1986) enacted policies focused on deregulation, privatization, and decentralization. These measures lowered barriers for services, trade, and capital—particularly financial capital—to move more freely, and made it easier to exploit labor. The policies of the Lubbers I cabinet marked the final break with the cabinets of the 1970s, which had still followed state-aid strategies and used public tax money to revitalize struggling national sectors. In line with the Single European Act, the Lubbers I cabinet propagated the motto "more market, less government" ("meer markt, minder overhead"), herewith legitimizing its strategies of deregulation, neoliberal re-regulation, and privatization.[34]

This governmental strategy, however, did not lead to a decrease in state involvement, but rather a transformation in its nature and scope. Instead of corporatist tendencies in the form of close cooperation between trade unions, business associations, and government, which marked Dutch post-war industrial policy, the Dutch government focused on a market-led recovery of national industries by "bringing wage, energy and environmental costs under control, improving the operation of labor markets, simplifying regulations and stimulating investment" from 1980 onwards.[35] These changes were particularly conducive to foreign investments in Dutch industries. Through increased shareholding in the Dutch paper industry West European, Scandinavian, Finnish, and North American corporations pursued consolidation strategies in order to strengthen their European market positions.[36] In effect, the Dutch paper industry was no longer 'Dutch' in terms of ownership as capital networks extended far beyond the Netherlands.

Furthermore, the Dutch government, with backing from employer associations, no longer considered the criticisms raised by labor unions, especially concerning significant cutbacks to the welfare state, to be pertinent or impactful. The alignment between the government and employer associations successfully undermined the impact of labor unions' concerns and limited their ability to effectively challenge the implementation of austerity measures.[37] Indeed, the Dutch government put in place a legal

[34] Den Hertog (2003, p. 47), Patel and Weisbrode (2013, p. 141), and Van Damme (2006, p. 8).

[35] Wolinetz (1989, p. 90).

[36] Bouwens (2003, p. 20).

[37] Wolinetz (1989, p. 79).

framework that primarily safeguarded the interests of corporate shareholders, rather than considering the broader spectrum of stakeholders, including workers and consumers.[38] Especially, the *vermarkting* (sell-out) of businesses and marketing strategies by the state played into the hands of the financial class fraction and succumbed businesses to the volatility of the same. In the case of the Dutch paper industry, the increase in foreign industrial investment commonly took place in the form of M&As, also referred to as brown-field investments. The new center-right cabinet considered economic concentration in the form of M&As a suitable strategy to distribute economic risks, profit from economies of scale and combat illegal forms of collaboration, such as cartels and other collusive practices, which were no longer tolerated and seen as a major crime against consumer-welfare.[39]

Particularly in the context of the Dutch paper industry, M&As were considered a solution to the periodic issue of overproduction, which continually haunted the industry. The cyclical sensitivity of the industry to overproduction is closely linked to its capital intensity.[40] Investments into production facilities are comparably high in the Dutch paper industry, but necessary for the expansion of production under rising demand.[41] Unable to coordinate their investments with precision, "[d]ifferent companies tended to invest at the same time, causing a discontinuous growth of production capacity, during which periods of overcapacity occurred".[42] Therefore, buying production capacity rather than building it became the guiding investment principle of the 1980s, herewith allowing for the maximization of profits by circumventing the industry's cyclical character.

The efforts of de- and re-regulation under the Lubbers I cabinet had an overall positive effect on the Dutch paper industry's profit growth during the 1980s. The deregulation of the public sector involved its privatization through selling state shares and listing state-run industries, such as the railway industry, the health sector, the energy sector, and telecommunication sector on stock markets. The privatization of large parts of the energy sector led to a considerable increase in the number of power

38 Horn and Vliegenthart (2010, pp. 60–63).

39 Bouwens and Dankers (2010, p. 760).

40 Bouwens (2003, p. 14).

41 Ibid.

42 Ibid.

plants. In turn, this caused a reduction in energy costs of about 50 percent for the Dutch paper industry, substantially contributing to the industry's profit growth in the 1980s.[43]

In addition, the government pursued re-regulation aimed at establishing new forms of collaboration among state agents, industrial companies, and research institutes to address the complexities inherent to innovation processes.[44] An illustrative example of this approach was the financial state support provided to the VNP, the national lobbying organization of the Dutch paper industry. This funding enabled the VNP to initiate several projects aimed at reducing energy costs, increasing the utilization of waste paper, and minimizing ecologically undesirable pollution within the Dutch paper industry.[45] These projects fostered collaboration with the *Nederlandse Organisatie voor Toegepast-Natuurwetenschappelijk Onderzoek* (TNO, Netherlands Organization for Applied Scientific Research), a sectoral organization operating under public law. The partnership between TNO and VNP resulted in the development of less risky and costly methods for the treatment of recovered paper, surpassing the profitability of previously available techniques. This cooperative effort exemplified the government's commitment to supporting innovation and finding sustainable solutions within the industry.[46]

A crucial factor contributing to the profit growth of the Dutch paper industry during the 1980s was the forced reduction in labor costs, which aligned with the broader trend of labor market flexibilization in the post-Fordist mode of regulation. Two ministerial reports from 1979 and 1980 that criticized Dutch industrial sectors for being outdated and underperforming are of the essence here. These reports sparked public concern, which in turn provided ample opportunity for the Lubbers I committee to implement their labor deregulation policies. Comprised of top government officials, independent experts, and board members of transnational corporations (TNCs), the cabinet's policies included labor flexibilization and wage restraints. These measures were intended to enhance the competitiveness of Dutch industries, including the paper industry,

[43] Bouwens (2003, p. 19) and Stellinga (2012, p. 29).

[44] Siraa (2016, p. 65).

[45] Bouwens (2004, p. 199).

[46] Ibid., p. 133.

by reducing labor costs. The implementation of these policies aligned with the broader post-Fordist approach, emphasizing the adaptability and flexibility of labor markets. While these measures contributed to the profitability of the Dutch paper industry, it is important to note that they also had implications for labor's ability to organize and negotiate for better working conditions and wages.[47]

The cabinet's official goal in restructuring governmental activity was to stimulate entrepreneurial activities and rise national market competitiveness by decreasing bureaucratic hurdles, improving efficiency, and lowering public expenditure.[48] However, the adopted Major Operations Program actually included (1) deregulation to "further market liberalization and competition", (2) de-nationalization to "economize on the state budget and to strengthen the market sector", and (3) privatization to "trim[...] the welfare state and establish[...] a smaller and more effective public sector".[49] Thus, the industrial downswing of that time was the basis for legitimizing the intensification of post-Fordist modes of regulation during the Lubbers I cabinet.

In 1990, the chairman of the earlier Lubbers I deregulation committee was appointed secretary-general of the Department of Economic Affairs under the coalition of Social Democrats and Christian Democrats—the Lubbers III cabinet (1989–1994)—heralding neoliberal re-regulation as a political priority.[50] In alignment with the European market integration, the Ministry of Economic Affairs redistributed governmental spending in favor of overall deregulation, increased labor market flexibility, and intensified competition policies. Shortly after, in 1994, the ministry implemented a new project called Market Operation, Deregulation and Quality of Legislation, which was not just aimed at legislative deregulation in favor of businesses, but gave businesses an actual right to a say in the re-regulation of, for example, environmental laws.[51]

These developments were closely intertwined with the transnationalization of finance and productive capital during post-Fordism. This rise

[47] Den Hertog (2003, p. 48).

[48] Hulsink and Schenk (1998, p. 246).

[49] Ibid., pp. 245–246.

[50] Den Hertog (2003, p. 52).

[51] Milieufocus (2008).

of TNCs represents the emergence of the transnational capital class fraction, consisting of "four main fractions: TNC executives, globalizing bureaucrats, globalizing politicians and professionals, and consumerist elites, including merchants and media".[52] TNCs "have become dominant forces in the transfer of capital, production, and technology in the global political economy", shaping industries, such as the Dutch paper industry, during post-Fordism.[53] Deregulation of cross-border transactions in the interest of TNCs increased steadily while large European and US-American corporations started dominating the global (financial) market.[54] By 1990, more than 40 percent of the Dutch industries were foreign-owned, and four of the top ten corporations in Europe continued to acquire production units in the Netherlands.[55]

"[C]ozy relationships between government, big business, labor [as in pillarized unions] and the financial community" benefitted the emergence of mega-mergers, such as KNP and Bührmann-Tetterode in 1993, and accelerated the process of transnationalization of ownership and production processes.[56] Executives of vastly growing Dutch-based transnationals, such as Phillips, Unilever, AkzoNobel, Heineken, and Royal Dutch Shell were actively involved in Dutch policy-making through chairing industrial policy advisory committees and, thus, constituted an increasingly powerful transnational class fraction.[57] They would suggest the implementation of business-friendly projects such as the *technolease scheme*, which allowed for "indirect state subsidy amounts" to enable big transnationals to financially exploit their "undepreciated know-how" at little financial risk.[58] By the end of the 1990s, a small number of transnational corporations, namely Stora-Enso, Norske Skog, Kappa Packaging, and SAPPI, who all belonged to the top 10 TNCs of Europe at that time, held a market share of up to 85 percent in the Dutch paper industry.[59]

[52] Carroll and Carson (2015, p. 71) and Sklair (1997).

[53] Overbeek and van der Pijl (1993, p. 260).

[54] Buch-Hansen and Wigger (2010, p. 33).

[55] Bouwens (2012, p. 204).

[56] Hulsink and Schenk (1998, p. 244).

[57] Sluyterman and Nieuwegracht (2004) and Sluyterman and Wubs (2014).

[58] Hulsink and Schenk (1998, p. 244).

[59] Bouwens (2012, pp. 204–205).

From 2000 onwards, the Dutch paper industry shrunk in size and profit as shareholders followed asset-stripping strategies to seek short-term profitability by purchasing plants and closing them down respectively. The disengagement of international investment capital from the Dutch paper industry and the ultimate moving of industrial activities abroad, namely the process of deindustrialization, left only four out of twenty-seven companies of the Dutch paper industry Dutch-owned at that time.[60] Even though the industry survived by staying located within the national borders of the Netherlands to a certain extent, deindustrialization remains the industry's number one threat until today. To "manag[e] the process of internationalization [...] in the hope of minimizing its harmful domestic repercussions and/or of securing maximum benefit to its own home-based transnational firms and banks" is the main goal of post-Fordist industrial policy and industry associations alike.[61] Thus, efforts of the Dutch government and the VNP to revitalize national industries by convincing foreign investment *to stay* in the Netherlands through promoting and improving the national conditions for manufacturing paper and board remain fundamental to the survival of the Dutch paper industry.

5.2 Network Cooperation

The historical conjuncture when the post-Fordist accumulation regime gained prominence in the Netherlands was influenced by the surge of neoliberal ideas advocating for a competition regime. Entailing the prohibition of cartels and the intensified enforcement of competition laws, the establishment of the *Nederlandse Mededingingsautoriteit* (NMa, The Dutch Competition Authority) in 1998 resembles the full blown arrival of the post-Fordist competition regime in the Dutch political economy. Dutch cartel legislation and, more specifically, the establishment of the NMa are essential for understanding *how* illegalized yet widespread practices of cooperation, like cartels, were substituted by network cooperation to foster knowledge sharing as well as product and process innovation, safeguarding the industry's survival during post-Fordism.

[60] Bouwens (2003, p. 10).

[61] Jessop and Sum (2006, p. 108).

During the early 1980s, there was a notable absence of prosecution against cartels, leading to the characterization of the Dutch economy as a 'cartel heaven'. However, as international attention grew toward numerous Dutch cartel cases, it triggered a gradual accumulation of pressure from the European Commission (EC) on the Dutch authorities regarding cartel prosecution at the national level.[62] These pressures stemmed from the implementation of an extremely lenient policy toward economic consolidation at the European Community level in 1990.[63] Even though early revisions of Dutch competition policies led to the registry of hundreds of cartels at the Ministry of Economic Affairs in the beginning of the 1990s, the national adaption of competition regulation was rather slow. It was not until 1994 that the Netherlands fully embraced EC cartel regulations by introducing a comprehensive ban on, price and market cartels, merger control, and anti-competitive market dominance abuses.[64] Nevertheless, hundreds of cartels persisted in operating illicitly, resulting in a substantial level of market concentration within the Dutch paper industry during the latter half of the 1990s.[65] Consequently, the reputation of the Netherlands as a haven for cartels endured, and the industrial supremacy of large incorporated manufacturing operations continued to expand.[66]

By 1998 the Competition Act, namely "[…] a series of anti-trust measures [and] a full-fledged framework of merger control provisions", was implemented and, subsequently, the NMa was founded under the Ministry for Economic Affairs.[67] As competition laws were enforced more rigorously, especially through the prosecution of collusive practices like price-fixing and market sharing, the end of the Dutch cartel haven was approaching. Active prosecution of cartels sincerely threatened Dutch industries' competitive advantages in a globalized market as collusive practices had been a common and tolerated form of cooperation in Dutch industries. In 1998, the at that time State Secretary for Economic Affairs van Rooy commented on the adaption of EC cartel

[62] Bouwens and Dankers (2014, p. 59) and Drahos (2001, p. 24).

[63] E.g. Wigger and Buch-Hansen (2014).

[64] Drahos (2001, p. 24).

[65] Bouwens and Dankers (2010, pp. 769–770).

[66] Ibid., p. 770.

[67] Hulsink and Schenk (1998, p. 248).

regulation as follows: "[...] Competition policy had to find a balance between combating cartels, on the one hand, and stimulating productive forms of cooperation, on the other".[68] In fact, the intensification of the competition regime and subsequent rise in prosecution of cartels generated an industrial policy whereby R&D type of cooperation sponsored by the state became more prominent. Hence, competition policy in the Netherlands during post-Fordism was always also *industrial* policy.

It is in this context that the *Kenniscentrum Papier en Karton* (KCPK, Paper and Board Knowledge Centre) emerged to facilitate cooperation within the Dutch paper industry: Since the prosecution of cartels as well as the intensification of merger legislations posed genuine threats to the industrial activities and economic performance of the Dutch paper industry, the KCPK became the Dutch paper industry's hub for network cooperation. Partially state-funded, the KCPK stresses the difference between cooperative projects for technological innovation among companies of the Dutch paper industry and actual cartel practices, which are now legally banned on national and EC-level due to the intensification of anti-collusive legislations.

The establishment of the KCPK in 1998 was, thus, more than convenient for the declining industry. Until that point, the companies of the Dutch paper industry occasionally cooperated in innovation projects with the TNO and the Wageningen University & Research (WUR). This close cooperation between publicly financed organizations on the one side, and private companies, on the other, laid the foundation for the KCPK to emerge. After extensive discussions between managers of the TNO, WUR, and the board members of the VNP, it was decided that the Dutch paper industry was in need of a knowledge center. In 1998, all three organizations (VNP, TNO, WUR) signed an agreement to finance the operations of the KCPK and its projects to one-third. This funding agreement was part of further long-term agreements signed between the VNP and the Ministry for Economic Affairs during the 1990s.[69]

Through the establishment of the KCPK, cooperation continued despite the rising suspicion toward collusive practices.[70] The incentive behind initiating the KCPK was to facilitate innovation projects between

[68] As cited in Drahos (2001, p. 371).

[69] Chappin et al. (2008, p. 1467).

[70] Bouwens and Dankers (2010, p. 770).

companies of the Dutch paper industry in order to provide joint knowledge outcomes under optimal use of resources. The KCPK did not and still does not provide its own laboratory for carrying out research. Instead, the KCPK manages research projects by distributing different research tasks to the actual mills or R&D centers of the participating companies. Herewith, the KCPK does not compare to research institutes for paper production innovation in other countries such as Germany, France, and Norway. These are, in contrast, privatized, corporate entities, which provide contracted services to paying customers.

If, for example, a paper company in Germany wishes to test a prototype for a more efficient use of fibers, but has no internal R&D laboratory or means to run these tests within the actual mill, they pay the respective research institute, in this case the *Papiertechnische Stiftung (Paper Technology Foundation)*, to run the tests for them. In the early 2000s, a comparable organizational restructuring was deliberated for the KCPK, proposing to place it under the authority of the WUR, essentially transforming it into a research institute without its previous independent status. As per one interviewee, this move could have jeopardized the KCPK's primary role in securing national and international funding for research initiatives within the Dutch paper industry. By opting to retain the initial organizational setup of the KCPK, the uninterrupted progression of research projects, primarily benefiting a select group of private companies and aiming to enhance the profitability of the Dutch paper industry, was made possible.

During the post-Fordist era in the Netherlands, public procurement strategies, including the one supporting the KCPK, were prevalent forms of state regulation in the industrial sector. State authorities adopted a reactive and less transparent approach to financially supporting collaborations through the KCPK. Instead of allocating public funds to industrial projects based on a clear strategy, they adhered to the funding agreement of the KCPK, which involved a one-third contribution from each party without a maximum threshold. Consequently, the KCPK initially secured funding from Dutch paper industry companies and then approached TNO and WUR to match this amount. Only after this matching process, the funds were allocated to one of the two project streams of the KCPK. This extensive funding availability led to a growth in staff numbers from an initial 12 to 30 by 2001, resulting in increased costs for maintaining the KCPK as an independent entity.

Organizations like the KCPK were potentially liable to fall into the category of competition-distorting (semi-)government organizations, according to EU competition rules.[71] The fact that the KCPK was established as a knowledge center for R&D projects, but did not and still does not feature its own R&D laboratory, could have potentially made it into a case of "[u]nfair competition from entities related to the government".[72] Yet, the NMa was not able to prosecute unfair competition practices, if conflicts of interest arose with its superordinate, the Ministry of Economic Affairs.[73] Due to its lack of independence, the NMa was not able to provide unobstructed prosecution of collusive practices. To be able to do so in the future, it needed to become an independent governing body.[74] Until that happened, it was "most important", as the Director General of the NMa was paraphrased in a 1999 OECD report on the role of competition policy in the Netherlands, "[...] that NMa be, and appear to be, independent in its decision-making. [To avoid] bureaucratic disputes and embarrassments, [...] it will be focusing now on cases it can uphold against a challenge in court".[75] Thus, it was a sole matter of time until the NMa became independent, possibly investigating projects facilitated by the KCPK as competition distorting (semi-)government organizations.[76]

The termination of the all-to-one-third funding agreement of the KCPK in 2004 was well-timed, since the coming independence of the NMa in 2005 was foreseeable by policymakers and industrial agents alike—also for the Dutch paper industry. Already from the late 1990s onwards, the later director of the KCPK maintained close contact with the senior policy advisor and project manager of the Ministry of Economic Affairs. Through such close state-industry relations important legislative changes, like the NMa's independence, were sure to be considered in the substantive restructuring of the KCPK. Initially, a steady stream of state funding was guaranteed until the all-to-one-third funding arrangement for the KCPK concluded in 2004. Additionally, the board of directors at VNP had ample time to reorganize the legal framework of the KCPK

[71] OECD (1999, p. 22).

[72] Ibid., p. 25.

[73] Konings et al. (2001, p. 3).

[74] Drahos (2001, p. 381)

[75] OECD (1999, p. 22).

[76] OECD (1999, p. 25).

to align with both European Commission-level and national competition regulations, while also prioritizing the industry's profitability through the promotion of collaboration.

In 2004, the board of directors of the VNP, which at that time was comprised of one-third of all Dutch paper companies' executive managers, decided to privatize the KCPK via a membership format. Hereby, the KCPK could persist in advancing the interests of the Dutch paper industry, all the while avoiding classification as a competition distorting (semi-)government entity. The privatization of the KCPK implied that 30 percent of all approved project costs were financed through a new state subsidy system. The remaining 70 percent of project costs were paid by the respective organizations participating in the project. This new legal construction of the KCPK aligned with statutes for industrial project subsidies, which had been passed already in 1996 and allowed different ministries, including the Ministry of Economic Affairs, to allocate subsidies to industrial innovation projects that were in line with national guidelines for competitiveness as well as ecological standards.[77]

The new legal structure of the KCPK was not unique. Similar public procurement-based R&D structures were replacing all sorts of public–private-partnerships in the Netherlands from 2000 onwards. This switch in post-Fordist industrial policy was legitimized by claiming "recognized public interest in certain kinds of advances".[78] Hence, contemporary public procurement strategies are frequently portrayed as considering the public's interests in terms of industrial innovation. In reality, these strategies function as a means to distribute public funds to companies with minimal oversight and bureaucratic administration, similar to the approach taken with the KCPK. Consequently, in the post-Fordist era, industrial policy shifted from state support in the form of co-funding, as seen with the KCPK until 2004, to state support under the guise of public interests, exemplified by the present state subsidy system.

As a non-profit organization, the KCPK does not qualify for those national and EU funds, which are exclusively available to for-profit entities, such as SMEs or corporate businesses. In 2004, the KCPK, thus, established a subsidiary, Bumaga BV, which runs under the accounting template of an industrial, for-profit company. Bumaga and the KCPK list

77 Staatsblad (1997).

78 Martin and Scott (2000, p. 440).

the same, eight employees on their websites.[79] The offices of the KCPK and Bumaga are both registered at the same address. Even though legally Bumaga is a private company and the KCPK is a foundation, the function and aim of both organizations concerning the Dutch paper industry are very similar, if not identical. On the webpage of Bumaga is stated that the company focuses on product and process innovations through "[…] project management, project support, government support, financial support, market explorations, market introductions and patents and licenses for new technologies".[80] Similarly, the KCPK aims at generating funding for the purposes of innovation, longevity, and capital growth for (companies of) the Dutch paper industry.[81]

Various funding options at both the national and EU levels cater to different types of applicants within the manufacturing industry. Some are exclusively for collaborative innovation ventures among corporate entities, while others are reserved for projects that include non-corporate entities like research institutes or foundations, such as the KCPK. Therefore, it is advantageous for the KCPK to seek funding as either a foundation or a private company. As a result, the KCPK tends to be more frequently engaged in process innovation projects, whereas Bumaga is more inclined toward product innovation projects.

The above explications show, how cooperation changed under the post-Fordist competition regime. The rising suspicion and actual prosecution of collusive practices through the NMa placed pressures on manufacturing industries, especially on already struggling ones, like the Dutch paper industry. In light of the changing relations of (re-)production, the influence of the competition regime, the increasing transnational ownership, and the ongoing deindustrialization in the post-Fordist era, novel modes of cooperation and industrial policies emerged. These enable the state to allocate public funds toward industrial projects. Fundamentally, recently established network organizations such as the KCPK and Bumaga are a proactive response to address the challenges confronted by manufacturing industries in the post-Fordist era. As the subsequent section illustrates, cooperative efforts to tackle global competition persist even beyond the scope of network organizations like the KCPK.

[79] Bumaga BV (2017) and KCPK (2017).

[80] Bumaga BV (2017).

[81] KCPK (2017).

Current forms of cooperative and competitive practices for the Dutch paper industry exist on several levels: The local, national, and the global level. Locally, cooperation between Dutch paper mills themselves and between Dutch paper mills and other industrial sites takes place regularly and partially in secret. Examples cover a vast spectrum of different forms of cooperation. They range from harmless lending of machinery to secret cooperative endeavors, in which companies abuse their powerful positions by imposing vertical constraints on suppliers' sales prices. For example, one interviewee admitted to have participated in a joint project with direct competitors for innovating end-products. They also explained that practices of jointly forcing suppliers to lower their prices are common in the industry. Other interviewees reported on less harmful and competition-stifling practices, such as lending machinery or expertise in case of emergencies between mills.

Nevertheless, also these 'harmless' practices are often secret as some of these mills belong to globally competing TNCs and their CEOs would not approve of such cooperation. Interestingly enough, all these forms of cooperation are based on individual relationships. Interviewees pointed to the importance of trust and respect for the success of project cooperation. One interviewee explained how manifold cooperation projects at their mill are not based on contracts, but solely on personal trust and respect. Another interviewee described how trust enables relationships to outlast actual projects and how trusting, personal relationships lead to sharing information outside of formal agreements.

Essential to these trusting and often personal relationships is that many of the employees of Dutch paper mills, especially in historically grown paper manufacturing regions such as Gelderland (overlapping with the once famous paper region of the Veluwe), have been working in this industry and partially the same mill for over forty years. The low turnover rate, also in the case of white-collar workers, is essential for these everyday forms of local cooperation. As one manager mentioned in an interview, they know each and everyone in the Dutch paper industry and in order to prevent cartel structures and personal cooperation, the whole industry personnel would have to be replaced as most of them maintain special, personal bonds. So even though CEOs officially do not encourage cooperation based on personal relations, an interviewee confirmed that "the cozy part [is] still there". Other interviewees agree, that, at times, these personal relations can negatively affect formal cooperation in innovation projects as "the click" determines whom to choose as a partner for specific projects and whom not to choose. Despite these personal preferences for cooperation partners, one interviewee states that if "the boss imposes on you to cooperate with someone, you just do it". Hence, personal relationships between paper mill employees, especially white-collar ones, are pivotal for both short-term solutions of manufacturing emergencies as well as abuse of powerful market positions. Yet, they can be detrimental to top-down initiated, formal cooperation in innovation projects.

Cooperation on the national level is often more formalized than on the local one. Examples are the public swimming pool Coldenhove, which gets the waste heat of the paper mill Coldenhove as well as other projects, in which paper mills and other organizations try to make use of the mill's waste heat with the support of the provincial government.[82] Another example is a shared wastewater treatment plant, which originated as a subsidiary of three paper mills in Gelderland (Mayr-Melnhof Eerbeek BV, DS Smith Paper De Hoop Mill, and Coldenhove Papier BV); its foundation was supported by the Surface Water Pollution Act in 1970.[83] Also the establishment of regional innovation parks to raise the efficiency in energy and water resource exploitation is exemplary of cooperation on the national level.[84]

In line herewith, government authorities and the VNP held the Target Group Negotiations to positively influence the competitive environment of the Dutch paper industry by measures ranging from the implementation of severe "anti-pollution standards to the subsidizing of cost-intensive restructuring programs".[85] With direct support of the state, the VNP focused on developing and improving energy-saving technologies, on finding new raw material sources, and on intensifying their relations with supply chain organizations to attract international investments.[86] Additionally, the Ministry of Economic Affairs and the VNP jointly signed a long-term agreement on three major industry topics: Sustainability (energy reduction), process and product innovation, and

82 Provincie Gelderland (2010) and Zwembad Coldenhove (2017).

83 Industriewater Eerbeek (2013).

84 Zeemeijer (2016, May 8).

85 Bouwens (2012, p. 206) and Chappin et al. (2008, p. 1467).

86 Bouwens (2012, p. 205).

networking to build vital connections with other national and international industries as well as municipalities, research facilities, and NGOs.[87] In alignment with such installations, the character and future vision of the Dutch paper industry was re-defined into prioritizing energy-saving innovations to stay competitive. Another example of a state-industry joint venture is the agreement between the government and a particular paper mill in the Netherlands to receive low-price timber. This timber was originally produced under governmental lead to manufacture *klompen* (traditional Dutch shoes) at the beginning of the twentieth century. An interviewee explained that when the market for *klompen* decreased drastically, the government was looking "for an industrial application for this forest and […] made a joint venture with this mill" in the early 1970s. This joint venture still runs today.

In summary, cooperative practices at the national level align with industrial policies that involve state-industry agreements aimed at fostering technological innovation to enhance the international competitiveness of the Dutch paper industry. Consequently, the significance of cooperation at the national level did not diminish in the post-Fordist era. In contrast, competition notably decreased at the national level, with most competitive activities occurring on a global scale within the Dutch paper industry. Both, the sustained significance of cooperation as well as the decline of competition on the national level are effects of the industry's transnationalization and subsequent deindustrialization during post-Fordism.

The transnationalization of capital, materializing in, among others, the mega-mergers of the 1990s, led to the domination of foreign TNCs above SMEs and family-owned domestic production units.[88] Due to this development, market shares of specific paper branches in the Netherlands are highly concentrated. In branches such as light-weight coated paper and newsprint manufacturing, the five largest producers covered a market share of up to 85 percent in the early 2000s.[89] By 2005, the remaining 27 paper mills in the Netherlands were owned by 17 companies, of which 16 belong to TNCs; in 2016, the number of paper mills further decreased

[87] Bouwens and Dankers (2014, p. 59).

[88] Bouwens (2012, p. 204).

[89] Ibid., p. 205.

to 21. The challenges arising from the transnationalization of production and ownership, namely the decreasing number of Dutch paper mills, actually reinforce the importance of the industry's national identity. This is why the national aspect of paper manufacturing is still relevant, despite the fact that the industry is almost entirely transnationalized in terms of ownership.

Industrial strategies of the VNP are geared toward developing, stressing, and strengthening a national identity to convince investment capital to stay within the spatial borders of the Dutch state. This is also visible in the continuous reference to the *Dutch paper industry* by employees of the industry or at industry events: The emphasis on national identity in papermaking and the continued emphasis on cooperation at the national level are only possible in the post-Fordist era, which is characterized by increased competition because very few of the remaining paper mills in the Netherlands directly compete in terms of consumer markets. In fact, interviewees confirmed that Dutch paper mills foremost compete with *foreign* paper mills of the same or other TNCs for market segments instead of one another. This is a common problem associated with the rise of TNCs in the global economy, as "[…] TNCs can deploy influence over their subsidiaries in ways less available to domestic firms, notably through 'coercive comparisons' between sites in different countries".[90] One of the interviewed managers indirectly referred to this dilemma and described a recent case of market consolidation in the Dutch paper industry, in which foreign investors bought paper mills in the Netherlands as well as in new-growth markets, to subsequently shut down the less profitable production site to decrease competition.

Such forms of intra-organizational competition, namely Dutch mills competing with foreign production sites of their main shareholding TNC that focus on similar consumer markets, are further infused by financialization. Liquid capital follows the harsh market carvings of asset-stripping strategies, private equitization, and debt-led risk investments.[91] These trends represent a big threat to the Dutch paper industry. Interviewees describes that the industry "get['s] eaten up by the world around [and] by bigger countries, [leaving] rising productivity [as] the only way to survive:

[90] Bélanger and Edwards (2006, p. 28).

[91] E.g. Harvey (2007, p. 161).

faster, more, cheaper". One interviewee, belonging to the older managerial generation of the Dutch paper industry, describes private equity capitalism as "absolute crazy", "extreme", and the source of "all the shit in the world", explaining that no one cares about the continuity of the mills anymore, but only about short-term profit maximization. Spurred by the threats of relocating production sites to new-growth markets in Asia and South-America, the decreasing number of Dutch paper mills goes hand in hand with a rise in niche-focused production.[92]

To conclude, the number one challenge for the Dutch paper industry during post-Fordism was to keep foreign investment located within the Netherlands. The answer of the dying Dutch paper industry to the threat of global competition under intensified anti-cartel legislation remains similar to what it has always been: Cooperation on the local and national levels. To not get busted by national and supranational anti-collusion bodies, the legitimate label for cooperation in post-Fordism is *network*. Networks, aimed at innovation, connect national paper producers with (inter-)national research centers, relevant government organizations, other national sectors of suppliers and buyers, and international paper producers. These networks do not only replace cartels in the Dutch paper industry but they balance the contradictory dynamics of competition-based cooperation. In times of anti-cartel legislation and European integration, new industrial policies serve the purpose of justifying such networks. Overall, cooperation remains an essential aspect of industrial survival in post-Fordism.

5.3 Circular Economies

During post-Fordism, technological innovation foremost revolves around a drastic increase in the speed and scale of papermaking machinery. In the Netherlands, the total number of paper and board machines decreased by 11 between 1993 and 2002.[93] Twenty machines, which produced less than 50,000 tons of paper per year, were shut down and nine machines, which produce more than 50,000 tons of paper per year, were newly

[92] Jäger and Springler (2015, p. 117), Jessop (2002, p. 99), and Jonker-Hoffrén (2013, p. 276).

[93] Bouwens (2004, p. 303).

established.[94] This rising speed and scale of paper machines placed the Netherlands as eighth of all 90 CEPI paper production states worldwide in terms of annual production output between 1993 and 2002.[95] Worldwide, smaller-scale paper machines were shut down in order to compensate for the implementation of large-scale ones; in 2003 the largest European paper machine was established in Belgium producing 400,000 tons of paper annually.[96]

In addition to increasing the speed and scale of machinery, the search for new raw materials and energy sources became fundamental to the survival of the Dutch paper industry during post-Fordism. Newly emerging paper producers in several South-East-Asian countries quickly started to dominate the global paper market due to cheaper access to raw materials and energy sources than elsewhere in the world. Thus, independence from importing raw materials continued to be essential for the profitability of the Dutch paper industry.[97] As a result, not only the already well-established development of paper recycling, but topics such as efficient energy usage (and, thus, energy reduction) as well as the re-usage of primary waste streams through integral chain management and resource circularity became the number one targets of state-industry cooperation during post-Fordism in the Netherlands.

Strategies such as integral chain management and resource circularity demand cooperation between suppliers, buyers, state agents, research facilities, and end-consumers. In the last two decades, ideas of finding cooperation partners to develop cradle-to-cradle systems and entire chains for reusing residuals became prevalent. According to interviewees, "you just can't do innovation alone" became a widely proclaimed principle in the Dutch paper industry. At the same time, rising suspicion toward collusive practices as well as competition-led market structures undermine cooperation for technological innovation in so far as interviewees admit that it is guided by the credo: "The first one who will build it, gets the profit". Consequently, cooperative endeavors for the sake of technological innovation are caught between the pressures of competition on

[94] Ibid.

[95] Ibid., p. 300.

[96] Ibid., p. 304.

[97] Ibid., pp. 304, 306.

the one side and the need for cooperation on the other.[98] It is in this context that the KCPK's focus on sustainability evolved as a solution to this complexity.

The focus of the KCPK on *sustainable* innovation is embedded in a wider agenda on the energy transition of the European Union as well as the Netherlands. At the beginning of 2000, the Ministry of Economic Affairs developed a national program for energy transition that covers multiple sectors as part of The Fourth Dutch National Environmental Policy Plan (NMP4).[99] Even though the Dutch paper industry is "bio-based", meaning that it is relatively sustainable as it uses renewable and recyclable raw materials, it is one of the largest consumers of fossil fuel in Europe.[100] The overall goal of the NMP4 was to decrease energy usage by 50 percent by 2020 and foster the transition to low-carbon energy usage.[101] Part of this policy plan is the Dutch program 'circular economy', initiated by the Ministry of Infrastructure and Environment in cooperation with the Ministry of Economic Affairs in 2016.[102] The circular economy program seeks to align industrial production cycles with resource circularity as found in natural ecosystems by fostering close cooperation between government agents and representatives from science, NGOs, and business.[103]

Circular economy, indeed, describes the minimization of using "the environment as a sink for residuals [and]—perhaps more importantly—[…] virgin materials for economic activity".[104] The business-friendly adaption of this buzzword in the Netherlands and elsewhere follows a contrary logic. As stated in the circular economy program of the Netherlands,

> [t]he circular economy also presents our country with plenty of (economic) opportunities. Innovation creates opportunities for existing businesses, for newcomers (start-ups), and for science. […] The circular economy can

[98] E.g. Sveiby et al. (2012).

[99] Pöyry Management Consulting Oy (2014).

[100] International Energy Agency (2022).

[101] Pöyry Management Consulting Oy (2014).

[102] The Ministry of Infrastructure and the Environment (2016).

[103] Ibid.

[104] Andersen (2007, p. 133).

> thus make a significant contribution to the future earning capacity of the Netherlands and Europe. The Netherlands has a good starting position to capitalise on these opportunities. [...] The Rabobank has estimated that a circular economy can lead to extra growth in GDP ranging from 1.5 billion euros (in a business-as-usual scenario) to 8.4 billion euros (in the most circular economic scenario).[105]

Discourses on circular economy and sustainability within and beyond the Dutch paper industry are predominantly guided by principles of profitability and non-fiscal economic advantages, which are to be realized through the ample application of new business models, new forms of labor exploitation, and efficient resource access as well as usage.[106] This is not surprising, since industrial policies continue to follow the basic capitalist principles of growth and profit in post-Fordism.[107] The circular economy program is based on the idea "[...] that virtually any social problem is subject to a technical and technological fix", ultimately succumbing ecological sustainability to economic sustainability.[108] Current environmental programs are, thus, tools to legitimately allocate public money to innovation projects, which ease the way for industries to stay competitive while implementing a minimum of national as well as supranational ecological standards.

Concretely, the Dutch practice of guiding national industries toward the sole reliance on renewable energy and bio-based raw materials is an instrument to keep Dutch industries internationally competitive. The subsidiary of the Ministry for Infrastructure and Environment called *Nederland circulair!* (Circular Netherlands) brings it to the point by stating that the existing possibilities for reusing and recycling raw materials only allow for manufacturing third-rate products, which in turn leads to the annihilation of millions of Euros every year. Ultimately, the Dutch circular economy program exemplifies how the role of politics has been "[...] reduced to finding the technical means to achieve goals (e.g.,

[105] The Ministry of Economic Affairs (2016, pp. 10–11).

[106] European Commission (2012), Ghisellini et al. (2016), and Lieder and Rashid (2016).

[107] Fisher (2010), Jackson (2009), and Smith (2016).

[108] Fisher (2010, p. 232).

economic growth) that in themselves are understood to lie outside the realm of politics".[109]

Also the KCPK, as the Dutch paper industry's innovation hub, is entrenched with the post-Fordist buzzwords of circular economy and sustainability. The circular economy program is a tool to increase the Dutch paper industry's economic sustainability, rather than facilitating its growing ecological sustainability.[110] In line herewith, the KCPK claims that most of its innovation projects lead to economic efficiency *and* environmental performance.[111] Hence, whatever is environmentally friendly needs to be economically profitable first. The primary focus of the KCPK is on technological innovation to drive the industry's international competitiveness and (transnational) capital growth, thus, aligns well with the Ministry's circular economy program.

Concluding, the agenda of economic sustainability underlies both the purpose and impact of the circular economy initiative by the state as well as the herewith-funded innovation projects as facilitated by network organizations such as the KCPK. Ultimately, the label of sustainability legitimizes the allocation of public funding to networks geared at industrial survival and economic sustainability.[112] Thus, in post-Fordism technological innovation first serves the purpose of enhancing companies' competitive advantage, profit maximization, and cost efficiency by improving production processes and fostering market-driven product specialization, and only then serves the purpose of increasing industries' ecological sustainability.[113] As long as technological innovation is propagated as the most important factor for economic growth, organizational strategies aimed at improving environmental sustainability will remain rooted in the capitalist rationality of granting primacy to profit and growth above all else.[114] In fact, ecological sustainability and circular

[109] Fisher (2007, p. 160).

[110] KCPK (2017).

[111] KCPK (2016, February).

[112] Methmann (2010, p. 349).

[113] Fisher (2007) and Kelly (1999).

[114] Phillips (2014, p. 443).

industries are at best side-products of transnational capital accumulation through market-driven product specialization and efficiency-driven process innovation under the agenda of economic sustainability.[115]

5.4 New Forms of Precarity

During post-Fordism, the transnationalization of production and ownership paired with the dynamics of financialization concurs with the decline in labor union density in the Netherlands and beyond.[116] Through a closer look at the reproduction of precarious working conditions alongside the rise of the managerial middle class, the racialized economy of the capitalist mode of production appears relevant to the erosion of working-class cohesion in post-Fordist Netherlands.

Even though the weakening of labor unions' bargaining power from 1980 onwards is undeniable, labor-capital relations in the Netherlands are often romanticized as a successful example of the *polder model*. The polder model describes the consensus-based culture between dominant agents of labor, capital, and state during early post-Fordism. Many scholars, politicians, and media outlets mystified the Wassenaar Agreement of 1982 as a consensus-based wage restraint, which induced job and economic growth, feeding into the ongoing polder model propaganda.[117] Starting with the Lubbers I cabinet in the 1980s and continuing past the Lubbers III cabinet in the 1990s, "competitiveness has been considered as the central key to economic and employment growth" and "wage restraint [as permitting] enhanced international competitiveness [...] thus creat[ing] growth in output and employment".[118] Heralding labor rights destruction as the only way to secure capital accumulation and, ironically, rising wealth for all, is quite essential to post-Fordist labor market deregulation and competition-centered discourses.

As a prime example, the Wassenaar discourse, which propagates a Dutch culture of consensus between labor, capital, and state, is a farce. First, the various Wassenaar 'agreements' of the 1980s and 1990s coincided with decreasing numbers in union membership, decreasing union

[115] McManus (1996), Perkins (2007), Shiva et al. (2014), and Smith (1996).

[116] Stockhammer et al. (2016, p. 1818).

[117] De Vries (2014) and Delsen (2002).

[118] Becker (2005, p. 1079) and Becker and Schwartz (2005, p. 14).

power, and the neoliberal co-option of unions.[119] Due to this power imbalance between the state, industry, and unions, one can hardly argue that these agreements were consensual.[120] Second, these negotiations were not unique to the Dutch context and so-called polder model, but also appeared in other national contexts around the time of post-Fordist labor flexibilization and neoliberal re-regulation.[121] Actually, the Wassenaar 'agreement' of 1982 was a non-consensual declaration of intent by the social democratic union (FNV) and Christian-democratic union (CNV) on a "'cost neutral' redistribution of work".[122] Instead of the expected reduction in daily working hours for full-time employees, the implementation of the Wassenaar 'agreement' induced drastic wage restraints and labor flexibilization.[123] These entailed an enormous increase in part-time, zero-hour, low-wage jobs for women, migrant, and juvenile employees.[124] In fact, the Netherlands is a champion of labor market flexibilization exhibiting the sixth highest percentage of flexible workers in Europe.[125] In 2015, 40 percent of the Dutch labor force worked in flex-time contracts, earning on average 35 percent less than people with fixed employment.[126] Since the beginning of the economic crises in 2008, such forms of occupation increased at a much higher rate than full-time employment.[127] Therefore, it is not particularly meaningful to the majority of workers in the Netherlands that Dutch employment protection legislation for full-time, regular jobs remains among the most generous in Europe.[128]

In the Dutch paper industry, this rise of post-Fordist precarity materialized in an increased workload per worker, alongside the continuity of hazardous working conditions, which include extreme temperatures and humidity or rapid changes thereof, high noise disturbances, constant

[119] Becker (2001, p. 463) and Bieler (2009, p. 233).

[120] Becker (2001, pp. 464–466).

[121] Becker (2005, p. 1079).

[122] Ibid., p. 1085.

[123] Oorschot (2004), Remery et al. (2002).

[124] Becker (2005, pp. 1081, 1087).

[125] Barbieri (2009, p. 5) and Industrial Safety & Hygiene News (2017, April 11).

[126] Barbieri (2009, p. 4) and De Beer and Verhulp (2017, p. 8).

[127] Industrial Safety & Hygiene News (2017, April 11).

[128] Becker (2005, pp. 1087–1088).

exposure to dangerous chemicals (such as asbestos or different sulfites) and physical stress.[129] The increased workload per worker is often misconstrued as an alleged rise of employee efficiency. It actually is caused by a decreasing number of workers per mill in order to reduce employment costs. Such strategies yielded an excessive decline in employment numbers during the 1980s and 1990s, hereby stimulating post-Fordist deindustrialization further. While a comparably large paper mill would count around 2000 machine operators in the 1960s, the number decreased to a total of 200 workers per mill in 2016.

During the 1990s and 2000s, strategies of downsizing based on advice from external consultancies were deployed to increase profits and lower the cost of labor extensively. Only recently, paper mill directors started to argue for an increase in workforce per mill as the profits a single worker accrues actually outgrew their employment cost. One interviewee explained the revenue maximization per employee as follows: "[W]orkers will make more profit than [we] have to pay for them". A rise in profit through increased worker's productivity is nothing more than the essence of precarity inherent to having ever less workers produce ever-more paper under hazardous working conditions.[130] During an interview one manager summarized this issue as follows: "So each worker is at ninety percent of their capacity, without having problems. So it means they don't even have time anymore to think [...]". This manager's assessment of the intense working conditions of paper machine operators in the Dutch paper industry is highly biased. Revealing the manager's refusal to acknowledge blue-collar working conditions in Dutch paper mills, they continue to legitimize profit-squeezing strategies such as downsizing. While rationalizing peoples' ability to cope with work as "capacity" and exploiting this "capacity" to its limit, the comment "not even having time anymore to think" exemplifies management strategies of revenue maximization at the cost of workers' welfare.

The second cause for the decreased employment numbers in the sector is the dire shortage of educated blue-collar workers. Under the Lubbers I cabinet and continuing into the 1990s, the education sector got re-regulated, marked by cuts of public funding and a focus on specialized

[129] Jungbauer et al. (2005), PKGV (2001), Torén, Hagberg, et al. (1996), Torén, Persson, et al. (1996).

[130] Jonker-Hoffrén (2011, p. 387), and Tregenna (2011, p. 3).

education as well as prioritization of higher education. As one interviewee explains,

> [t]hey killed this system at the end of the 80s, beginning 90s, and then they turn over to the overall education and MAVO and things. But they now, they are starting again with the leerling system. [...] And this is something, the whole industry is facing. That we slept for 10, 15 years for education because poor education means also you have to invest time and resources.

In other words, post-Fordist trends of deindustrialization are inter-related with lower vocational education rates. For these reasons, the average worker's age in the Dutch paper industry is around fifty years and one interviewee explains that "the biggest problem in this industry [...] is keeping the knowledge, transferring the knowledge within the company before they go on retirement". One could think that the old-workers' knowledge might be redundant in a fully automated paper mill, but it is not. A high rate of machine-stoppage mostly due to paper ripping, approximately once per day in most Dutch paper mills, requires the machine operators to manually intervene with the machine beyond, as one interviewee plays it down, just "pushing a button". A widely spread alternative to retirement-induced knowledge gaps is to buy-in the relevant knowledge from experts. These practices are costly and not always efficient because such experts, usually studied engineers, do not operate paper machines on a daily manufacturing basis.

In sum, the concurrence of the decrease in unionism and the rise in productivity occurs primarily at the cost of the workers. The demand for ever higher output of ever fewer employees under hazardous conditions reproduces workers' precarity in post-Fordism. This development is further underpinned by the relocation of production sites to emerging markets (deindustrialization) and the re-regulation of the educational sector, which focuses less on manufacturing personnel and more on the service sector and knowledge workers. In line herewith, the following section explores the development of the managerial middle-class in relation to capital and labor during post-Fordism.

The emergence of the managerial middle class and its manifestation in the Dutch paper industry dates back to the 1950s and 1960s.[131]

[131] Kriesi (1989, p. 1088).

In the early 1980s, the Dutch paper industry, like many other industries worldwide, witnessed a drastic increase in managerial positions. This class fraction's growth is an essential facet of post-Fordist labor-capital relations.[132] The managerial middle-class is also referred to as the professional-managerial class (PMC),[133] new petty bourgeoisie,[134] or white-collar workers.[135] It "consist[s] of salaried mental workers who do not own the means of production and whose major function in the social division of labor may be described broadly as the reproduction of capitalist culture and capitalist class relations".[136] Herewith, managers take a unique position in post-Fordist labor-capital relations.

Corporate managers are neither fully in power of deciding their own working conditions, including their salary scale, their working hours, their liabilities, and tasks, nor dependent upon organizing their labor force in the form of unions to actually negotiate their working conditions. Instead, they inhabit a position that lies outside the union-negotiated tariff for different sectors. In the Dutch paper industry, the specific circumstances of managers are described as follows by an interviewee:

> First of all because [...] the competition between the companies for hiring educated people is just higher, so you need more freedom [to negotiate your working conditions], the second thing is, the working hours are quite long. Sometimes you stay fourteen, fifteen hours because there are some problems, you also are on call sometimes, considering that your salary actually doesn't increase that much. [...] So basically, the tariff doesn't apply to me [and] we have direct negotiations.

The flexibilization of working conditions also affects the managerial class fraction, as eight-hour working days are often exceeded. Due to the rise of modern technology, managers are expected to be available throughout the day and always in case of an emergency. In contrast to blue-collar workers, managers are in the position to negotiate their salaries and schedule their working days autonomously, often through—as one interviewee calls it—"gentleman agreements". Next to their unique bargaining positions

132 Ibid.

133 Ehrenreich and Ehrenreich (1979).

134 Poulantzas and Hunt (1975).

135 Burrell (2002) and Hyman and Price (2016).

136 Ehrenreich and Ehrenreich (1979, p. 13).

within organizations, the managerial middle class also exhibits powers concerning negotiations about future investments and legal boundaries of making (paper) business. This power simultaneously depicts their responsibility to maintain and increase the profitability of the business in the interest of the shareholders. Hence, negotiations between transnational shareholders and corporate managers are an integral part of post-Fordism since their interests do not always align.

In the Dutch paper industry, this new petty bourgeoisie of well-paid, mostly white, men in suits faces contradictory expectations, which are tied to their role within the corporate mill. Paper mill managers have to negotiate their role as representatives of the particular TNC, which owns the paper mill, and their role as representatives of the *Dutch paper industry*. One manager describes this tension as follows:

> Actually, I'm not loyal to [name of TNC], [...] [i]n my personal opinion, I'm loyal to this mill. Because here we are fighting for not being closed down in some future. We are small. So, we are fighting and the head quarter is deciding they have [number of] mills, we are too small in profit. So, I'm loyal to this factory. I'm part of [TNC], but my first loyalty goes to this location.

In the case of the Dutch paper industry, paper mill managers operate under the continuous tension of keeping 'their' mills profitable, while fearing the ever-present possibility that the shareholders shut down the mill to diminish competition, decrease overcapacity, or follow asset-stripping strategies.

This threat became an actual, dominant trend after 2000, when TNCs started to move production units abroad, close down machines, or dismantle entire plants.[137] In essence, this leaves managers in a contradictory situation: Identifying with the Dutch paper industry and fearing the shutdown of 'their' mill while at the same time executing their shareholders' will, which solely adheres to rising profits and growth, instead of the national and local importance of the mill. Dutch paper mill managers confirmed in the interviews, that they feel a need to protect the national paper industry from their untamable, corporate employers. The identification of white-collar employees with the *Dutch* paper industry is, thus, essential to the survival of the industry.

[137] Bouwens (2012, p. 205).

Despite the industry's transnationalization, the low turnover rates and decade-long careers are still the norm. In fact, most TNCs follow corporate strategies, which adhere to the fact that subsidiaries are embedded in distinct national cultures, which do not necessarily align with their respective culture.[138] Thus, they continue to employ local managers and engineers, instead of replacing the white-collar staff with foreign professionals. Overall, mill managers face an arduous and complex future when it comes to ensuring the survival of the Dutch paper industry by convincing foreign capital to stay.[139] This also explains, why the VNP stresses the importance of the *Dutch* paper industry, even though the industry is almost entirely foreign-owned and its managers (including most of the board members of the VNP) are employed by foreign TNCs.

The rise of the managerial middle-class as a post-Fordist abscess of labor-capital relations posits the continuity of distinct elements of the colonial past, namely the racial segregation of the economy as manifested in post-Fordist labor-capital relations. Thus, underlying the rise of the managerial middle class is the continuation of an "imperial racial economy, with its gendered, sexualized, and classed intersections".[140] During the era of post-Fordism, manufacturing work in first-wave industrialized countries, such as the Netherlands, was relocated to women, migrant, and juvenile workers. Intertwined with this shift in employment was the post-World War II migration wave, "consist[ing] of three major groups: postcolonial migrants from the (former) empire, labor migrants from the circum-Mediterranean area and recently from Eastern Europe, and refugees from a variety of countries in Africa, Asia, Latin America, and the Middle East".[141] Relocating manufacturing jobs to the outsider (within)[142] in order to exploit and appropriate their workforce, changed the racialized economy in so far as industrial blue-collar jobs were no longer limited to white people.[143] The current role migrant workers play

138 Schneider (1988).

139 Bouwens (2012, p. 205).

140 Wekker (2016, p. 2).

141 Ibid., p. 6.

142 A concept used and developed by Gloria Wekker, describing the silenced experiences of people, which are part of a society, yet invisibilised and thus not seen as such (Wekker 2001, 2016).

143 Brown (2003, pp. 70–71).

in manufacturing industries of first-wave industrialized countries depicts the racialized character of the post-Fordist accumulation regime as it describes (forced) migration,[144] allowing corporations to exploit labor at ever cheaper prices due to low (or non) labor regulations for women, migrant, and juvenile workers.[145]

In addition, white supremacist racism continues to underpin the creation and attempted legitimation of the managerial identity.[146] Current racist justifications of the oppressive idea(l)s and practices of management date all the way back to early forms of Taylorism during slavery.[147] A statement by one of the managers in the Dutch paper industry reproduces the inherently racialized character of the white, male, highly educated managerial middle-class identity:

> We have some areas, where the academic level doesn't have to be quite high, you are more there to operate three buttons basically, so we have a high percentage of Turkish people, or with Turkish origins. [...] Ehm, for the low, ehm, academic level jobs, well you hire basically not illiterate people, but people not with a high school degree. [...] I think, I'm not sure, if they are officially Dutch, we more define it as people, who speak Dutch. [...] I mean, if you are able to communicate, it's fine for us. Not if you are in a position, where you have to co-ordinate, but if you understand, what we are telling you to do, it's fine.

This quote has to be interpreted as embedded in and reproducing the racialized economy, to which the Dutch paper industry is no exception. Until the partial decolonization after World War II, manufacturing jobs in the metropoles remained predominantly white until they were handed down to migrant workers in first-wave industrialized countries.[148] Due to the interplay between the European integration and the national labor deregulation policies, the Dutch paper industry exhibited growing numbers of migrating blue-collar workers from 1980 onwards.[149] They

[144] For further discussions on the nexus of forced migration, globalization and (neo-) colonialism please see: Angathangelou and Ling (2003).

[145] Castles (2002, p. 1144), and e.g. Castles et al. (1998).

[146] Cooke (2003, p. 1905).

[147] Ibid.

[148] Brown (2003, pp. 70–71).

[149] Castles et al. (1998, p. 72).

mostly migrated from countries such as Turkey, Poland, Bulgaria, and Rumania, and, once arrived in the Netherlands, faced short-term working contracts, low social security standards, and little to no support from unions.[150]

For the respective manager to establish a causal relationship between the assumingly simple task of pushing three buttons and the number of (assumingly) Turkish people working in a mill is racist. It is equally racist to link the notion of Turkish people to the idea of illiteracy and it is discriminatory to indirectly equalize not having a high school degree with illiteracy (despite the fact that shaming illiterate people by giving illiteracy a negative connotation is already discriminatory in itself). Overtly racist statements like the one cited above and the concomitant racist atmosphere pertaining not only in Dutch manufacturing industries but as a matter of fact in the Dutch societal context at large, are supported by an epistemology of ignorance.[151] In fact, similar to how Dutch colonialism was mystified as a 'golden' age, the racist structures analyzed here are eagerly kept secret through the white Dutch self-representation as "a small but ethically just nation that has something special to offer to the world".[152] This ignorance enables white supremacy to continue its reign by forcefully defending non-knowing, innocence, and the absence of race and racism from Dutch society.[153] In a similar vein, it is claimed that racism, if at all, only takes place among the working class; it is said to surely not be part of the (managerial) middle-class in Dutch societies.[154]

Of course, this is not surprising as management (as a practice and scientific discipline) is essentially rooted in the colonial project.[155] Techniques of Othering reflect the racist roots of management as established by esteemed scholars, who ignored, how slavery and, thus, white supremacist racism was and continues to be incremental to management as practice and science.[156] Exemplary,

[150] Ibid.

[151] Sullivan and Tuana (2007).

[152] Wekker (2016, pp. 4, 13).

[153] Ibid., p. 18.

[154] Ibid.

[155] Frenkel and Shenhav (2006, p. 7).

[156] Cooke (2003).

> Ferrero's studies, which contrasted the stereotypical image of the lazy barbarian with that of the self-controlled, industrious and civilized European, were accorded scientific credence and were instrumental in shaping the worldview of important theoreticians of organization and management vis-a-vis the non-western 'other'.[157]

Othering is a strategy commonly deployed to build a dominant identity and maintain racialized, oppressive structures. In science Othering is reinforced through institutionalized discourses, which "[...] are predicated upon a colonial tradition where non-white people are seen as exotic Others who require Western scientific investigation and classification".[158] The discursive use of Othering installs and legitimizes violent and oppressive structures such as patriarchy, nationality, borders, law, Western[159] science, and various forms of state and corporate violence, which are all based on building an identity in demarcation to the Other, who does not belong.[160] Essentially, Othering within and beyond science employs the following strategy: "If you want to be equal to us, then don't talk about differences; but if you are different from us, then you are not equal".[161] In the Dutch context,

> [p]ersistently, an innocent, fragile, emancipated white Dutch self is constructed versus a guilty, uncivilized, barbaric other, which in the past decades has been symbolized mostly by the Islamic other, but at different times in the recent past blacks (i.e., Afro-Surinamese, Antilleans, and Moluccans) have occupied that position.[162]

The above-cited interview respondent, titled deputy manager, is himself a migrant worker from a West-Middle-European country, who enjoyed a middle-class upbringing and an international engineering education at

157 Frenkel and Shenhav (2006, p. 9).

158 Liu (2017, p. 50).

159 I use Western here and in later passages of the text as a descriptor for the mythical and fictional nature of "the West [a]s an ideological construction deployed in the asymmetrical power relations between colonizer and colonized" (Westwood & Jack, 2007, p. 248).

160 Shiva et al. (2014, p. 255) and Mignolo (2012).

161 Prins (2002).

162 Wekker (2016, p. 15).

different paper mills worldwide. His father, also a paper engineer, worked at many different paper mills worldwide and actively supported his son's career. Essentially, this vita resembles how the patriarchal and (neo-) colonial character of the Dutch paper industry, especially concerning the industry's management level, is still based on 'gentleman agreements'. Furthermore, assuming to inhabit a position, in which 'he the manager' is legitimated to define another person's identity ("we more define it as people, who speak Dutch"), ultimately resembles the racist, authoritarian, patriarchal, and imperial core of the managerial middle-class. Even though 'he the manager' tries to hide this core by using the seemingly legitimizing plural "we", he assumes a judge-like authority for himself (or in his words for "us") by devaluing the Other. Additionally, the racist, authoritarian, and discriminatory nature of the manager's last sentence is worth noting, as he clearly sees himself in a role, in which he tells the Others, what to do, while they have to follow his order.

To conclude from the above explications: The reproduction of racist discourses and material practices is marked by the establishment and rise of the managerial middle-class and its inherently racialized character. Furthermore, the managerial middle-class is embedded in and reproductive of a society, which claims to be blind toward and, hence, perpetuates racialized realities. Overall, current (forced) migration and subsequent work displacement to oppressed groups in post-Fordism goes hand in hand with neoliberal restructuring processes yielding "fiscal austerity, the liberalization of trade agreements, a reduction of import tariffs, and wide-ranging cuts to public expenditure".[163] By these means, racial inequality and labor market discrimination are also characteristic of the Dutch paper industry, exploiting so-called low-skilled migrant workers, who work under low social security standards and flex-time contracts, while facing racist oppression. Overall, the rise of the managerial middle-class is essential for reproducing the oppressive structures inherent to capitalism by augmenting the hierarchy between capital and labor under the guise of extrapolating capital growth during post-Fordism.

[163] May et al. (2008, pp. 63–66).

5.5 Rebutting State Retrenchment

The role of the state in post-Fordism is conditioned by and its policies adjusted to the international competition regime, which stresses the importance of national competitive advantages based on "territory, population, built environment, social institutions and economic agents".[164] The showcased active and passive governmental steering and restructuring of economic affairs in the context of the Dutch paper industry from 1980 onwards refutes the thesis that the state retrenches from regulative involvement during the post-Fordist accumulation regime. Instead, the above analysis shows how the state's regulative involvement within and beyond the Dutch paper industry simply changes throughout time but does not cease. Despite the contradictory propagation of free market principles and rising competition, the cooperation between government authorities and managers of the Dutch paper industry clearly traces manifold instances of industrial policy, which benefitted the profitability of the industry. State strategies, such as deregulation of market structures allowing for a widespread transnationalization of ownership, neoliberal re-regulative activities in the form of closer cooperation between public and private sectors for the purpose of R&D cooperation, processes of privatizing public sectors, active as well as passive financial support for industry demands, and flexibilization of labor, all pursued the goal of increasing the international competitiveness of Dutch industries. The governmental response to post-Fordist threats of deindustrialization, the moving of production units abroad, and subsequent shut-down of paper mills in the Netherlands was to bolster industrial and, hence, capital growth through setting "strategic targets for flexible accumulation, continuous innovation and the promotion of the overall structural competitiveness of the national economy".[165] Herewith, the post-Fordist state continued to play a fundamental role in the restructuring of the politico-economic environments its national industries are embedded in.

During post-Fordism the number one goal for the Dutch paper industry was to soften the negative repercussions of rising deindustrialization and to put a stop to the moving of production capacities to third-wave industrialized countries.[166] In the Netherlands, among

[164] Jessop (2002, p. 96).

[165] Jessop (1994, p. 268).

[166] Kurosawa and Hashino (2017, pp. 22ff.).

others, the competition regime is promoted as the only way to make economic growth attainable; keeping fully transnationalized industries located within the respective national borders is the number one goal of post-Fordist states. With the full-blown arrival of the competition regime in the Netherlands, namely the establishment of the NMa, new forms of cooperation between companies as well as between the public and private sectors became fundamental. Foremost targeting R&D cooperation, industry-specific knowledge and innovation hubs, like the KCPK, started sprouting. Focusing on the national identity of *Dutch* paper-making became pivotal for propagating R&D cooperation despite the growing dominance of the competition regime. With the goal to foster cooperation, the KCPK established local, regional, and national level innovation projects in order to strengthen the industry's competitiveness on the global level. Partially funded by (supra-)national institutions, the KCPK, as a daughter organization of the VNP, aimed to increase the attractiveness for paper manufacturing businesses to stay in the Netherlands.

The opposing forces of cooperation during heightened competition manifest in the demand for certain forms of industrial cooperation, i.e., networks aimed at innovation, and the disdain for other forms of industrial cooperation, i.e., cartel structures. In fact, network forms of industrial R&D-focused cooperation are politically legitimized under fashionable notions, such as circular economy. Innovative production processes, technology, and products are propagated as sustainable solutions to ecological problems. Sustainability is the new buzzword under which the repercussions of deindustrialization are said to be tamable by re-orienting manufacturing alongside market demands and niche products. Embedded in a discourse of ecological sustainability, technological innovation actually prioritizes the need for capitalist profit accumulation and economic growth during post-Fordism. In this time of hyper-competition, cooperation has never been about increasing industries' ecological sustainability.[167] Overall, the innovation discourse led in the name of ecological sustainability more generally and in the context of the Netherlands in the name of the circular economy program more specifically, disregards any positive aspects of collusive practices, which

[167] D'aveni (2010) and Langevoort (2002).

could help diminish the negative impacts competition-induced overproduction and its subsequent increase in waste have on our natural and social environments.

Also within the final dimension of labor-capital relations, one can find tensions. New forms of precarity arose in first-wave industrialized countries, while harsh and exploitative labor was relocated to other geographical as well as demographic areas. Deplorable working conditions in newly industrialized countries provide the basis for a culture of ever-growing consumption in first-wave industrialized countries. Also in European manufacturing industries, which officially adapted supranational safety and production standards, new forms of precarity arose. These include flex-time, low-paid jobs, and decreasing social security standards, especially for women, migrant, and juvenile workers. While automation and computerization in manufacturing industries seem to have decreased the harsh working conditions of blue-collar workers, time-efficiency and unethical productivity measures are actually increasing the daily pressures on workers in the Dutch paper industry. Not only does the dominance of capital over labor remain, but this power relation is further augmented by the rise of the managerial middle-class. As this class mostly consists of white, male, middle-class "Westerners"[168] in the Dutch paper industry, its patriarchal and (neo)colonial character is sustained. Yet, also managers are confronted with certain contradictions in the post-Fordist accumulation regime. These relate to the fact that managers are both, internationally mobile representatives of big TNCs as well as locally placed managers, who, in the case of the Dutch paper industry, identify more with the respective industrial site than their transnational corporate employer.

Furthermore, the rise of the managerial class fraction lends impetus to the above-described contradictions of cooperation in the dominant competition regime. Local and national cooperative efforts, including short-term, daily activities as well as long-term, formalized projects, are strongly reliant on personal relations between managers (including various management levels, from executive manager to technical engineer) in the Dutch paper industry. Globally, competition prevails over cooperation as

[168] I use the word *Westerner* as a critical notion to point to the discourse of Otherness. The term is established by a dominant group, in this case self-identified Westerners, whereby assuming to embody the norm and having a valued identity in contrast to the Other, "that is defined by its faults, devalued and susceptible to discrimination": Staszak (2008).

most Dutch paper mills do not compete on a national level, but more so with their globally spread sister mills. Currently, the biggest threat for Dutch paper mills is that their corporate shareholders shut down production as a consequence of following asset-stripping or consolidation strategies. Cooperation and technological innovation, as fostered by the Dutch state authorities, is thus designed to convince industrial investment capital to stay within the national borders of the Netherlands.

References

Altvater, E. (2009). Postneoliberalism or postcapitalism? The failure of neoliberalism in the financial market crisis. *Development Dialogue, 51*(1), 73–88.

Altvater, E., & Mahnkopf, B. (1997). The world market unbound. *Review of International Political Economy, 4*(3), 448–471.

Andersen, M. S. (2007). An introductory note on the environmental economics of the circular economy. *Sustainability Science, 2*(1), 133–140.

Angathangelou, A. M., & Ling, L. H. M. (2003). Desire industries: Sex trafficking, UN peacekeeping, and the neo-liberal world order. *The Brown Journal of World Affairs, 10*(1), 133–148.

Barbieri, P. (2009). Flexible employment and inequality in Europe. *European Sociological Review, 25*(6), 621–628.

Becker, U. (2001). 'Miracle' by consensus? Consensualism and dominance in Dutch employment development. *Economic and Industrial Democracy, 22*(4), 453–483.

Becker, U. (2005). An example of competitive corporatism? The Dutch political economy 1983–2004 in critical examination. *Journal of European Public Policy, 12*(6), 1078–1102.

Becker, U., & Schwartz, H. (2005). *Employment miracles: A Critical comparison of the Dutch, Scandinavian, Swiss, Australian and Irish cases versus Germany and the US*. Amsterdam University Press.

Bélanger, J., & Edwards, P. (2006). Towards a political economy framework: TNCs as national and global players. In A. Ferner, J. Quintanilla, & C. Sánchez-Runde (Eds.), *Multinationals, institutions and the construction of transnational practices* (pp. 24–52). Palgrave Macmillan.

Bieler, A. (2009). Trade Union resistance to neoliberal restructuring. In B. van Apeldoorn, L. Horn, & J. Drahokoupil (Eds.), *Contradictions and limits of neoliberal European governance* (pp. 232–249). Palgrave Macmillan.

Bouwens, B. (2003). *The internationalization of the Dutch paper and board industry*. Unpublished paper presented at the BHC/EBHA Meeting, Lowell, MA, USA.

Bouwens, B. (2004). *Op papier gesteld: de geschiedenis van de Nederlandse papier- en kartonindustrie in de twintigste eeuw*. Boom.
Bouwens, B. (2012). The paper and board industry in the Netherlands, 1800–2000. In J.-A. Lamberg, J. Ojala, M. Peltoniemi, & T. Särkkä (Eds.), *The evolution of global paper industry 1800–2050* (pp. 191–209). Springer.
Bouwens, B., & Dankers, J. (2010). The invisible handshake: Cartelization in the Netherlands, 1930–2000. *Business History Review, 84*(4), 751–771.
Bouwens, B., & Dankers, J. (2014). From paradise to hell: Post-war cartel policy and practice in the Netherlands. *ESKA: Entreprises et Histoire, 3*(76), 58–71.
Bratsis, P. (2014). Political corruption in the age of transnational capitalism. *Historical Materialism, 22*(1), 105–128.
Brown, M. K. (2003). *Whitewashing race: The myth of a color-blind society*. University of California Press.
Buch-Hansen, H., & Wigger, A. (2010). Revisiting 50 years of market-making: The neoliberal transformation of European competition policy. *Review of International Political Economy, 17*(1), 20–44.
Bumaga BV. (2017). *About us*. Retrieved March 4, 2017, from http://www.bumaga.nl/about-us
Burrell, G. (2002). Twentieth-century quadrilles aristocracy, owners, managers, and professionals. *International Studies of Management & Organization, 32*(2), 25–50.
Carroll, W. K., & Carson, C. (2015). Forging a new hegemony? The role of transnational policy groups in the network and discourses of global corporate governance. *Journal of World-Systems Research, 9*(1), 67–102.
Castles, S. (2002). Migration and community formation under conditions of globalization. *International Migration Review, 36*(4), 1143–1168.
Castles, S., De Haas, H., & Miller, M. J. (1998). *The age of migration: International population movements in the modern world*. Palgrave Macmillan.
Chappin, M. M., Hekkert, M. P., Meeus, M. T. H., & Vermeulen, W. J. V. (2008). The intermediary role of an industry association in policy-making processes: The case of the Dutch paper and board industry. *Journal of Cleaner Production, 16*(14), 1462–1473.
Cooke, B. (2003). The denial of slavery in management studies. *Journal of Management Studies, 40*(8), 1895–1918.
D'aveni, R. A. (2010). *Hypercompetition*. The Free Press.
De Beer, P., & Verhulp, E. (2017). *Dertig vragen en antwoorden over flexibel werk*. University of Amsterdam.
De Vries, J. (2014). The Netherlands and the polder model: Questioning the polder model concept. *BMGN-Low Countries Historical Review, 129*(1), 99–111.

Delsen, L. (2002). *Exit polder model? Socioeconomic changes in the Netherlands*. Greenwood Publishing Group.

Den Hertog, J. A. (2003). *Public and private interests in regulation: Essays in the law and economics of regulation*. Utrecht University.

Drahos, M. (2001). Convergence of national competition laws. *The cases of Germany, Austria, and the Netherlands*. National Policy Adaption of the 19999 ESCA Sixth Biennial International Conference in Pittsburgh, Pennsylvania, USA, 3–6 June 1999.

Ehrenreich, B., & Ehrenreich, J. (1979). The professional-managerial class. In P. Walker (Ed.), *Between labor and capital* (pp. 5–45). South End Press.

European Commission. (2012). *Manifesto for a resource-efficient Europe*. https://ec.europa.eu/commission/presscorner/detail/en/MEMO_12_989

Fisher, E. (2007). Upgrading market legitimation: Revisiting Habermas's technology as ideology in neoliberal times. *Fast Capitalism, 2*(2), 159–171.

Fisher, E. (2010). Contemporary technology discourse and the legitimation of capitalism. *European Journal of Social Theory, 13*(2), 229–252.

Frenkel, M., & Shenhav, Y. (2006). From Binarism back to hybridity: A postcolonial reading of management and organization studies. *Organization Studies, 27*(6), 855–876.

Friedman, A. L. (2000). Microregulation and Post-Fordism: Critique and development of regulation theory. *New Political Economy, 5*(1), 59–76.

Ghisellini, P., Cialani, C., & Ulgiati, S. (2016). A review on circular economy: The expected transition to a balanced interplay of environmental and economic systems. *Journal of Cleaner Production, 114*, 11–32.

Harvey, D. (2007). *A brief history of neoliberalism*. Oxford University Press.

Helleiner, E. (1996). *States and the reemergence of global finance: From Bretton Woods to the 1990s*. Cornell University Press.

Horn, L. C., & Vliegenthart, A. (2010). Corporate governance in Nederland - hoe bedrijven handelswaar werden. In J. Blom (Ed.), *De Kredietcrisis - Een politiek-economisch perspectief* (pp. 59–78). Amsterdam University Press.

Hulsink, W., & Schenk, H. (1998). Privatisation and deregulation in the Netherlands. In D. Parker (Ed.), *Privatisation in the European Union: Theory and policy perspectives* (pp. 242–257). Routledge.

Hutton, W. (2008). The world we're in. Hachette Digital.

Hyman, R., & Price, R. (2016). *The new working class?: White-collar workers and their organizations—A reader*. Palgrave Macmillan.

Industrial Safety & Hygiene News (ISHN). (2017, April 11). *Report from Europe: Big increase in flexible workers in Netherlands. Gig economy uptick started during financial crisis of 2007–08*. ISHN. https://www.ishn.com/articles/106239-big-increase-in-flexible-workers-in-netherlands

Industriewater Eerbeek. (2013). *Industriewater Eerbeek - Sustainability Report*.

International Energy Agency. (2022). *Pulp and Paper*. IEA. Retrieved June 27, 2023, from https://www.iea.org/reports/pulp-and-paper
Jackson, T. (2009). *Prosperity without growth?: The transition to a sustainable economy*. Sustainable Development Commission.
Jäger, J., & Springler, E. (2015). *Asymmetric crisis in Europe and possible futures: Critical political economy and post-Keynesian perspectives*. Routledge.
Jessop, B. (1993). Towards a Schumpeterian workfare state? Preliminary remarks on post-Fordist political economy. *Studies in Political Economy, 40*(1), 7–39.
Jessop, B. (1994). Post-Fordism and the State. In A. Amin (Ed.), *Post-Fordism: A reader* (pp. 251–279). Blackwell.
Jessop, B. (2002). *The future of the capitalist state*. Polity Press.
Jessop, B., & Sum, N.-L. (2006). *Beyond the regulation approach—Putting capitalist economies in their place*. Edward Elgar Publishing.
Jonker-Hoffrén, P. (2011). Whose constituency? Representativeness of the Finnish Paper Workers' Union, innovation strategies, and outsourcing in the Finnish paper industry between 1980 and 2008. *Industrial Relations Journal, 42*(4), 375–391.
Jonker-Hoffrén, P. (2013). The influence of internal union factions on union renewal strategies: The case of the Finnish Paper Workers' Union. *Industrial Relations Journal, 44*(3), 276–295.
Jungbauer, F. H. W., Lensen, G. J., Groothoff, J. W., & Coenraads, P. J. (2005). Skin disease in paper mill workers. *Occupational Medicine, 55*(2), 109–112.
KCPK. (2016, February). *Valorization opportunities side-streams PBI*. Presentation at the Circular Economy conference of the Dutch paper and board industry, Arnhem, Gelderland, The Netherlands.
KCPK. (2017). *Algemeen*. Retrieved March 4, 2017, from http://www.kcpk.nl/algemeen
Kelly, K. (1999). *New rules for the new economy: 10 radical strategies for a connected world*. Penguin.
Koch, M. (2004). European reactions to global challenges: The cases of the Netherlands, the UK, Sweden, Spain, and Germany. *Journal of International & Comparative Social Welfare, 20*(1), 15–22.
Konings, J., Van Cayseele, P., & Warzynski, F. (2001). The dynamics of industrial markups in two small open economies: Does national competition policy matter? *International Journal of Industrial Organization, 19*(5), 841–859.
Kriesi, H. (1989). New social movements and the new class in the Netherlands. *American Journal of Sociology, 94*(5), 1078–1116.
Krippner, G. R. (2005). The financialization of the American economy. *Socio-Economic Review, 3*(2), 173–208.
Kurosawa, T., & Hashino, T. (2017). Three markets and three types of competitiveness: Pulp and paper industry. In B. Bouwens, P.-Y. Donzé, & T. Kurosawa

(Eds.), *Industries and global competition: A history of business beyond borders*, pp. 235–259. Routledge.

Lacher, H. (1999). Embedded liberalism, disembedded markets: Reconceptualising the Pax Americana. *New Political Economy, 4*(3), 343–360.

Langevoort, D. C. (2002). The organizational psychology of hyper-competition: Corporate irresponsibility and the lessons of Enron. *The George Washington Law Review, 70*(5–6), 968–975.

Lieder, M., & Rashid, A. (2016). Towards circular economy implementation: A comprehensive review in context of manufacturing industry. *Journal of Cleaner Production, 115*(March), 36–51.

Liu, H. (2017). Redeeming difference in CMS through anti-racist feminisms. In A. Pullen, N. Harding, & M. Phillips (Eds.), *Feminists and queer theorists debate the future of critical management studies* (pp. 39–56). Emerald Group Publishing Limited.

Lundan, S. M., & Mirza, H. (2011). TNC evolution and the emerging investment development paradigm. *Transnational Corporations, 19*(2), 29–52.

Martin, S., & Scott, J. T. (2000). The nature of innovation market failure and the design of public support for private innovation. *Research Policy, 29*(4–5), 437–447.

May, J., Datta, K., Evans, Y., Herbert, J., McIlwaine, C., & Wills, J. (2008). Travelling neoliberalism: Polish and Ghanaian migrant workers in London. In A. Smith, A. Stenning, & K. Willis (Eds.), *Social justice and neoliberalism: Global perspectives* (pp. 61–89). Zed Books.

McManus, P. (1996). Contested terrains: Politics, stories and discourses of sustainability. *Environmental Politics, 5*(1), 48–73.

Methmann, C. P. (2010). 'Climate protection' as empty signifier: A discourse theoretical perspective on climate mainstreaming in world politics. *Millennium, 39*(2), 345–372.

Mignolo, W. (2012). *Local histories/global designs: Coloniality, subaltern knowledges, and border thinking*. Princeton University Press.

Milieufocus. (2008). *Marktwerking, deregulering en wetgevingskwaliteit* (MDW). http://www.milieufocus.nl/factsheets/m/marktwerking-deregulering-en-wetgevingskwaliteit-mdw.html

The Ministry of Infrastructure and the Environment. (2016). *Circular economy in the Netherlands by 2050*. https://circulareconomy.europa.eu/platform/sites/default/files/17037circulaireeconomie_en.pdf

OECD. (1999). *Netherlands—The role of competition policy in regulatory reform*. https://www.oecd.org/daf/competition/2497317.pdf

Oorschot, W. V. (2004). Flexible work and flexicurity policies in the Netherlands. Trends and experiences. *Transfer: European Review of Labour and Research, 10*(2), 208–225.

Overbeek, H. (1993). *Restructuring hegemony in the global political economy: The rise of transnational neo-liberalism in the 1980s*. Routledge.
Overbeek, H., Van Apeldoorn, B., & Nölke, A. (Eds.). (2007). *The transnational politics of corporate governance regulation*. Routledge.
Overbeek, H., & van der Pijl, K. (1993). Restructuring capital and restructuring hegemony: Neo-liberalism and the unmaking of the post-war order. In H. W. Overbeek (Ed.), *Restructuring hegemony in the global political economy: The rise of transnational neo-liberalism in the 1980s* (pp. 1–27). Routledge.
Panitch, L., & Konings, M. (2009). Myths of neoliberal deregulation. *New Left Review, 57*(May/June), 67–83.
Patel, K. K., & Weisbrode, K. (2013). *European integration and the Atlantic community in the 1980s*. Cambridge University Press.
Perkins, P. E. (2007). Feminist ecological economics and sustainability. *Journal of Bioeconomics, 9*(3), 227–244.
Phillips, M. (2014). Re-writing corporate environmentalism: Ecofeminism, corporeality and the language of feeling. *Gender, Work & Organization, 21*(5), 443–458.
PKGV. (2001). *Sectorale Arbocatalogus voor de PKGV-industrie*. http://www.verbondpk.nl/arbocatalogus/
Poulantzas, N., & Hunt, A. (1975). *The new petty bourgeoisie. The new working class? White-collar workers and their organizations*. Springer.
Pöyry Management Consulting Oy. (2014). *The Dutch transition success story*. Retrieved June 22, 2023, from https://silo.tips/queue/the-dutch-transition-success-story?&queue_id=-1&v=1687440001&u=MTk1LjM3LjY0LjI1Mw==
Prins, B. (2002). The nerve to break taboos: New realism in the Dutch discourse on multiculturalism. *Journal of International Migration, 3*(3–4), 363–379.
Provincie Gelderland. (2010). *Quickscan Gelderse papierindustrie - Duurzamer door samenwerken*. https://www.dutchbiorefinerycluster.nl/download/108/documenten/Klimaat_Quickscan_Brochure.pdf
Remery, C., van Doorne-Huiskes, A., & Schippers, J. (2002). Labour market flexibility in the Netherlands: Looking for winners and losers. *Work, Employment and Society, 16*(3), 477–495.
Schneider, S. C. (1988). National vs. corporate culture: Implications for human resource management. *Human Resource Management, 27*(2), 231–246.
Shiva, V., Salleh, A., & Mies, M. (2014). *Ecofeminism* (2nd ed.). Zed Books.
Siraa, T. (2016). *Economische Zaken: een ministerie in historisch perspectief*. Elsevier.
Sklair, L. (1997). Social movements for global capitalism: The transnational capitalist class in action. *Review of International Political Economy, 4*(3), 514–538.

Sluyterman, K., & Nieuwegracht, K. (2004). *Ambitious strategies: Dutch multinationals between 1950 and 2000*. Retrieved September 11, 2023, from https://bintproject.nl/textfiles/2004_sluyterman.pdf

Sluyterman, K., & Wubs, B. (2014). Multinationals as agents of change. In K. Sluyterman (Ed.), *Varieties of capitalism and business history: The Dutch case* (Vol. 28, pp. 156–182). Taylor & Francis.

Smith, R. (2016). *Green capitalism: The god that failed*. College Publications.

Smith, R. J. (1996). Sustainability and the rationalisation of the environment. *Environmental Politics, 5*(1), 25–47.

Staatsblad. (1997). *Besluit subsidies economie, ecologie en technologie*. https://wetten.overheid.nl/BWBR0008443/2003-07-16

Staszak, J.-F. (2008). *Other/otherness*. International Encyclopedia of Human Geography. Elsevier.

Stellinga, B. (2012). *Dertig jaar privatisering, verzelfstandiging en marktwerking*. Amsterdam University Press.

Stockhammer, E., Durand, C., & List, L. (2016). European growth models and working class restructuring: An international post-Keynesian political economy perspective. *Environment and Planning A, 48*(9), 1804–1828.

Sullivan, S., & Tuana, N. (2007). *Race and epistemologies of ignorance*. SUNY Press.

Sveiby, K.-E., Gripenberg, P., & Segercrantz, B. (2012). *Challenging the innovation paradigm*. London, UK: Routledge.

Torén, K., Hagberg, S., & Westberg, H. (1996). Health effects of working in pulp and paper mills: Exposure, obstructive airways diseases, hypersensitivity reactions, and cardiovascular diseases. *American Journal of Industrial Medicine, 29*(2), 111–122.

Torén, K., Persson, B., & Wingren, G. (1996). Health effects of working in pulp and paper mills: Malignant diseases. *American Journal of Industrial Medicine, 29*(2), 123–130.

Tregenna, F. (2011). *Manufacturing productivity, deindustrialization and reindustrialization*. United Nations University (UNU)—World Institute for Development Economics Research (WIDER) (Working Paper, n. 2011/57). https://www.wider.unu.edu/publication/manufacturing-productivity-deindustrialization-and-reindustrialization

Van Apeldoorn, B., & Horn, L. (2007). The transformation of corporate governance regulation in the EU. In H. Overbeek, B. Van Apeldoorn, & A. Nölke (Eds.), *The transnational politics of corporate governance regulation* (pp. 99–119). Routledge.

Van Damme, E. E. C. (2006). Privatisering, deregulering en marktwerking. *Elementair, 5*(3), 8–10.

Varga, J. J. (2013). Breaking the heartland: Creating the precariat in the US lower rust belt. *Global Discourse, 3*(3–4), 430–446.

Watkins, S. (2005). Continental tremors. *New Left Review, 33*(May–June), 5–21.

Wekker, G. (2001). Of mimic men and unruly women: Family, sexuality and gender. In R. Hoefte & P. Meel (Eds.), *Suriname in the twentieth-century* (pp. 174–197). Ian Randle Publishers.

Wekker, G. (2016). *White innocence: Paradoxes of colonialism and race*. Duke University Press.

Westwood, R. I., & Jack, G. (2007). Manifesto for a postcolonial international business and management studies: A provocation. *Critical Perspectives on International Business, 3*(3), 246–265.

Wigger, A. (2015). Enhancing 'competitiveness' in response to the European crisis. In J. Jäger & E. Springler (Eds.), *Asymmetric crisis in Europe and possible futures: Critical political economy and post-Keynesian perspectives* (pp. 114–130). Routledge.

Wigger, A., & Buch-Hansen, H. (2014). Explaining (missing) regulatory paradigm shifts: EU competition regulation in times of economic crisis. *New Political Economy, 19*(1), 113–137.

Wolinetz, S. B. (1989). Socio-economic bargaining in the Netherlands: Redefining the post-war policy coalition. *West European Politics, 12*(1), 79–98.

Zeemeijer, I. (2016, May 8). *Nederland barst van de innovatievalleys, clusters en delta's*. Het Financieele Dagblad. https://fd.nl/economie-politiek/1150694/nederland-barst-van-de-innovatievalleys-clusters-en-delta-s

Zwembad Coldenhove. (2017). *Zwembad Coldenhove - Historie*. https://www.zwembadcoldenhove.nl/historie.php

CHAPTER 6

Networked Capitalism

Abstract The concluding chapter of the book consolidates the insights derived from each preceding chapter, emphasizing the fundamental importance of the state's engagement, labor conflicts, cooperative and competitive frameworks, and technological advancements in ensuring the survival of the Dutch paper industry. Despite being frequently scrutinized individually, these factors collectively serve as crucial elements in upholding the industry's resilience. The chapter concludes by looking ahead to the primary challenges anticipated for the industry and proposing sustainable trajectories for its future.

Keywords Industrial policy · Dutch capitalism · Dutch monarchic liberalism · Fordism · Post-Fordism

> While networked global capitalism is often aestheticized as smooth, continuous, and homogeneous, this system in fact requires asymmetry and discontinuity.
>
> After Globalism Writing Group (2018, p. 25)

To revisit the original question that shaped this book, which explores how the Dutch paper industry has weathered politico-economic shifts and managed to survive since its establishment in the late sixteenth century, the historicization of the Dutch paper industry reveals that various types of networks have played a crucial role in sustaining its

M. E. Ehrich, *The Dutch Paper Industry from 1580 to the Present*,
Palgrave Studies in Economic History,
https://doi.org/10.1007/978-3-031-54324-1_6

longevity across four distinct phases of capitalism. These networks have facilitated the industry's ability to adapt to technological advancements, address and mitigate labor uprisings, navigate competitive dynamics by fostering collaboration within the industry and with the government, and advocate for the interests of key industrial stakeholders at both national and international levels. The historicization highlights the significance of the state's involvement, labor struggles, cooperative and competitive structures, and technological innovations in safeguarding the survival of the Dutch paper industry. These factors, often examined in isolation, collectively play vital roles in maintaining the industry's resilience.

Particularly during the seventeenth and eighteenth centuries, when the Dutch Republic was a major economic and trading power, the Dutch paper industry held significant importance both globally and nationally. The Dutch capital, Amsterdam, emerged as a major center for paper production and trade. The industry benefitted from the country's access to waterways, including the River Amstel, which provided a convenient means of transporting raw materials and finished products. Dutch paper mills were concentrated in regions such as Gelderland and North Holland, where water-powered stamping mills proliferated and were known for their technological advancements. Aside from the ample water resources, the industry's growth can also be attributed to the Dutch Republic's expansive colonial trade networks, which facilitated the supply of raw materials, such as rags, as well as actual investment capital. The presence of a skilled local workforce propelled the industry even further. By requiring tasks such as collecting rags, operating machinery, and distributing paper, the sector created job opportunities, thereby fostering local economies in the process.

As the book showcases, labor is an essential component of surplus value accumulation in capitalism and thus crucial for the survival of the Dutch paper industry. The historical analysis presented here demonstrates the inseparable connection between labor-capital relations and the industry's endurance throughout all phases of capitalism. These relations are primarily characterized by the continuous process of proletarianization. During Dutch capitalism (1580–1815), the mechanization of production brought significant changes to the organization and structure of labor. As workers increasingly relied on urban wage labor, they began to organize themselves in associations advocating against war, exploitative work conditions, and colonialism, as a means of pushing back against capital-owners. The industrialization of production during the era of

Dutch monarchic liberalism (1815–1914) heightened this phenomenon, resulting in the assimilation of even the final remaining peasants in the Veluwe, who had previously participated in papermaking as a supplementary occupation, into the modern proletariat. Despite attempts by former peasants to regain independence through self-employment in straw-board cooperatives, these initiatives were ultimately undermined by the close networks among gentlemanly farmers, gentlemanly capitalists, and state authorities during the period of Dutch monarchic liberalism. These historical developments highlight the struggles and challenges faced by laborers and their attempts to resist exploitation, as well as the complex dynamics between different capitalist class fractions and the state throughout the various phases of capitalism.

With the end of World War II and the rebuilding of Europe through Marshall Plan aid, the criminalization of unionization was re-assessed during Fordism (1914–1980).[1] Even though unions were no longer criminalized, they nevertheless remained subordinated to the interests of the state and capital in the so-called pillarization of the Netherlands. In fact, strong employers' associations (e.g., chamber of commerce) and the rise of Taylorist working procedures excelled the weakening of (communist) workers' organizations further. During post-Fordism (1980 until now) labor-capital relations changed as new forms of precarity arose in first-wave industrialized countries, while harsh and exploitative labor was relocated to other geographical as well as demographic areas. In post-Fordism managers face the contradictory dynamics of being both internationally mobile representatives of big TNCs as well as locally placed managers, who, in the case of the Dutch paper industry, identify more with the respective industrial site than their transnational corporate employer. Subsequently, current forms of cooperation span 'old-men networks' within the Dutch paper industry, helping each other through more or less competition stiffening practices.

These competition-stiffening practices date back to early forms of cartel practices, which marked the Dutch paper industry's survival for centuries. Early contracts of correspondence secured capital accumulation for the merchant-capitalist class fraction during Dutch capitalism (1580–1815). Such forms of cartel structures, also labeled fire insurances, were cooperative ventures between Zaansian paper makers. Completely independent of

[1] Vickers (1998, p. 39).

the paper production in the hinterland, Veluwian peasants had no interest in usury and practices of lending money, instead relying on local support structures between families. The early forms of cartel structures in the Zaanstreek slowly led to a stark concentration of the industry during Dutch monarchic liberalism (1815–1914). Soon, institutionalized cartel structures secured the industrialization of Zaansian papermaking, in turn bringing Veluwian paper makers to start cooperating in cartel structures themselves in order to maintain handmade Dutch paper traditions.

With the rising internationalization and concentration of the Dutch paper industry during Fordism (1945–1980), cartels became a dominant form of cooperation, decreasing competition to a minimum. Paper makers in the Netherlands tried to impede the decreasing importance of manufacturing industries in comparison to the rising importance of service industries for the Dutch economy by initiating intensified cooperation. In fact, cartels remained the most viable route to safeguard Dutch industrial growth more generally and the growth of the Dutch paper industry more specifically, making the Netherlands internationally known as a cartel heaven during Fordism.

During post-Fordism (1980 until now), the decreasing importance of manufacturing industries reaches its zenith in the process of deindustrialization. To soften the negative repercussions of rising deindustrialization and to put a stop to the moving of production capacities to newly industrialized countries, Dutch authorities promote the competition regime as the only way to keep economic growth attainable and fully transnationalized industries located within the national borders. With the full-blown arrival of the post-Fordist competition regime in the Netherlands, namely with the establishment of the *Nederlandse Mededingingsautoriteit* (NMa, The Dutch Competition Authority), new forms of cooperation between companies as well as between public and private sectors became fundamental. These manifest in the contradictory dynamics of demanding specific forms of cooperation, i.e., networks aimed at innovation, and disdaining other forms of cooperation, i.e., cartel structures. As a result of foremost targeting R&D cooperation, industry-specific knowledge, and innovation hubs sprouted, such as the *Kenniscentrum Papier en Karton* (KCPK, Paper and Board Knowledge Centre). With the goal to foster cooperation, the KCPK established local, regional, and national level innovation projects in order to strengthen the industry's competitiveness on the global level. Therefore, a focus on national identity of *Dutch* papermaking became pivotal, propagating R&D cooperation among agents of national manufacturing industries. Partially funded by (supra-) national institutions, organizations like the KCPK, itself a daughter organization of the *Koninklijke Vereniging van Nederlandse Papier- en Kartonfabrieken* (VNP—The Royal Association of Dutch Paper and Board Mills), aim to increase the attractiveness for foreign direct investment to stay located within the Netherlands. In case of the Dutch paper industry, this entails the promotion of cooperation to increase technological innovation and thus keep Dutch paper manufacturing businesses profitable investment outlets.

Also technological advancements mattered substantially to the survival of the Dutch paper industry. In fact, the early international fame of Dutch papermaking was based on its famous invention, the *Hollander* beater. Only possible through financial cooperative structures this invention changed the tide between the now mechanized production of paper in the Zaanstreek and the continuation of handmade paper production in the Veluwe during Dutch capitalism (1580–1815). Cost-intensive innovations like the *Hollander* beater also determined the second shift in the history of (Dutch) papermaking. The comparably late implementation of the steam-run paper machine in the Netherlands during Dutch monarchic liberalism (1815–1914) once again reshaped the landscape of Dutch paper production, albeit to its detriment. Compared to its foreign competitors, Dutch papermaking remained small-scale. Only with strong state support did the two biggest players of the industry (KNP and van Gelder Zonen) manage to catch up with foreign developments, allowing for the production of 'endless' paper and the use of innovative, cheaply available raw material for board production, such as straw. By 1890, handmade paper production using cloth and based on the traditional *hamerbak* had completely stopped.

As a matter of fact, the following phase of Fordism (1945–1980) is marked by an increased search for new raw material sources to secure the independence of the Dutch paper industry from costly imports. Subsequently, the third major technological advancement—the innovative use of recycled waste paper—once more turns the tide for Dutch papermaking and increases its international competitiveness, yet not close to its former fame. This technological innovation was possible through networked cooperation between agents of the Dutch paper industry and state authorities. During post-Fordism (1980 until now), these networks were substituted by new forms of industrial R&D focused cooperation among agents of the Dutch paper industry as well as with state agents. Politically legitimized under fashionable notions, such as circular economy, innovative technology, production processes, and market-oriented products are propagated as sustainable solutions to ecological problems. Sustainability is the new buzzword under which the repercussions of deindustrialization are said to be tamable by re-orienting manufacturing along market demands and niche products. Hereby, technological innovation is embedded in a discourse on ecological sustainability, while actually prioritizing the need for capitalist profit accumulation and economic growth. Yet, in the post-Fordist time of hyper-competition, cooperation has never been about increasing industries' ecological sustainability.[2] Overall, the innovation discourse led in the name of ecological sustainability more generally and in the context of the Netherlands in the name of the circular economy program specifically, disregards any positive aspects of collusive practices, such as the prevention of overproduction and subsequent decrease in waste, which could help diminish competition-led exploitation of natural resources and labor.

In contrast to the belief that the role of the state has been minimized in the post-Fordist era, the findings of this book present a different narrative. The research demonstrates in meticulous detail, that the Dutch government, with varying degrees of involvement, has consistently played an important role to the survival of the Dutch paper industry throughout all phases of capitalism. Notably, during all phases of capitalism, state authorities have maintained close relationships with the dominant capitalist class fraction. In the era of Dutch capitalism (1580–1815), these connections were primarily established with the merchant-capitalist class fraction, which facilitated the flow of colonial monetary and resource capital into national manufacturing industries. However, it is important to note that these ties also prioritized favorable trading policies, which did not always align with the interests of manufacturers.

During the period of Dutch monarchic liberalism (1815–1914), the ties between the state and the Dutch paper industry underwent a partial transformation. These ties were replaced by connections with a new emerging capitalist class fraction known as the gentlemanly capitalists. Together, they embarked on a collaborative endeavor called 'Building Industria', which aimed to advance the interests of the manufacturing

[2] D'aveni (2010) and Langevoort (2002).

elite. However, it is important to note that this did not necessarily align with the interests of Veluwian papermakers with their family-owned mills and handmade paper production. The gentlemanly capitalist class fraction established close networking relationships with the monarchs initially and later with the liberal cabinet. Leveraging these close ties with state agents, the two largest paper and board manufacturing companies in the Netherlands at that time, KNP and van Gelder Zonen, exerted significant influence. As a result, they managed to economically overpower the Veluwian paper makers, compelling them to transition from handmade to mechanized paper production.

During the Fordism phase (1914–1980), the role of the state in actively restructuring national sectors became more pronounced. A closely knit network known as the 'old boys network', comprising influential political and business figures, played a significant role in securing support from the Dutch state to rejuvenate national industries. The state actively supported initiatives such as the exploration of innovative raw materials, offering substantial subsidies through joint ventures, overseeing sector restructuring and corporatization efforts, and exhibiting leniency toward cartel practices with minimal prosecution. Additionally, the state facilitated consolidation strategies and mergers and acquisitions within the Dutch paper industry and other national industries. These measures aimed to revitalize and strengthen the industrial landscape under the guidance of the state and under the influence of the 'old boys network'.

During the post-Fordism phase (1980 until now), the close networking ties between state authorities and capitalist class fractions appear to diminish. However, despite the propagation of free market principles and increased competition, the significance of state-industry relations for the survival of the Dutch paper industry remains evident. This book highlights numerous instances of industrial policy during post-Fordism that have contributed to the profitability of industries, including the Dutch paper industry. In fact, the state's initiation of market deregulation facilitated the transnationalization of ownership, attracting foreign direct investments into the Dutch paper industry. The overarching goal of enhancing the international competitiveness of Dutch industries, including the paper industry, primarily served the interests of the growing transnational capitalist class fraction. However, this increasing dominance also had adverse effects, such as the financialization of manufacturing industries and the relocation of production capacities to economically advantageous, third-wave industrialized countries.

In response to these challenges, the Dutch government implemented strategic objectives of neoliberal de- and re-regulation. These activities included closer collaboration between the public and private sectors to promote R&D cooperation, the privatization of public sectors, active and passive financial support for industry demands, and labor flexibility measures. These actions aimed to address the threats faced by the Dutch paper industry and other sectors within the framework of neoliberal policies.

Throughout the history of the Dutch paper industry, networks served as a means to increase capital accumulation by negotiating the interests of capitalist class fractions. It is crucial to acknowledge that the state plays a fundamental role in these negotiations across all phases of capitalism. The state's involvement encompasses initiatives to fund cooperative efforts in technological innovation for the development of new capital investment opportunities. Furthermore, the state implements policies focused on fostering competition to establish a so-called "level playing field", as articulated by former President of the European Commission, Juncker.[3] Also a main driver of labor flexibilization, it is important not to underestimate the role of the state throughout all phases of capitalism, even during periods characterized by pronounced and full-fledged (neo)liberal tendencies. The state's influence and engagement remain significant in shaping and supporting industries, including the Dutch paper industry, in various aspects of their development and operations.

In alignment with the recent revival of national and EU-level industrial policy, the historicization of the Dutch paper industry showcases the turnaround of the role of the state in market intervention signaling industrial policy to "be rising like a phoenix from its ashes".[4] Through the implementation of Juncker's agenda, which sought to "devalue labor, enhance competition, and lower corporate taxes" in order to amplify industrial growth and corporate profits in Europe, the industry once again confronts heightened competition and an increasing divide between capital and labor.[5]

To counteract these trends, the Dutch paper industry could shift its focus toward "horizontal and democratic solidarity economy initiatives"

[3] Juncker (2019, February 5).

[4] Wigger (2019, p. 353).

[5] Ibid.

that prioritize the interests of various stakeholders, including workers, citizens, indigenous cultures, activists, and environmental organizations, in the production of paper and utilization of natural resources.[6] In line with this approach, workers must reorganize and fortify their positions outside of the roles traditionally defined by unions within a capitalist framework. More inclusive organizations, encompassing not only workers but also individuals with a vested interest in the conditions of paper and board production, should emerge outside the dominant class fractions to form a potent opposition to capitalist interests.

Regarding technological progress, the industry could explore avenues beyond the confines of the capitalist paradigm, potentially leading to truly innovative methods of producing essential goods within a framework of solidarity. Achieving this would necessitate a global reevaluation of the incremental belief that competition inherently leads to growth. This reevaluation should consider natural resources as belonging to the collective, and recognize the Earth as a common resource for all inhabitants—humans, animals, and plants alike.

References

After Globalism Writing Group. (2018). Rerouting capillaries: Networked territories of condensation. *Social Text, 36*(1/134), 25–30.

D'aveni, R. A. (2010). *Hypercompetition*. The Free Press.

Juncker, J.-C. (2019, February 5). *Keynote speech by President Juncker at the EU industry days 2019*. Retrieved September 11, 2023, from http://europa.eu/rapid/press-release_SPEECH-19-870_en.htm

Langevoort, D. C. (2002). The organizational psychology of hyper-competition: Corporate irresponsibility and the lessons of Enron. *The George Washington Law Review, 70*(5–6), 968–975.

Pochet, P. (2016). From social pacts to productivity boards. *Intereconomics, 51*(6), 321–324.

Vickers, R. (1998). *Manipulating hegemony: British labour and the Marshall plan* (Doctoral dissertation, University of Warwick).

Wigger, A. (2019). The new EU industrial policy: Authoritarian neoliberal structural adjustment and the case for alternatives. *Globalizations, 16*(3), 353–369.

[6] Ibid.

Index

M. E. Ehrich, *The Dutch Paper Industry from 1580 to the Present*,
Palgrave Studies in Economic History,
https://doi.org/10.1007/978-3-031-54324-1

SPRINGER NATURE

GPSR Compliance

The European Union's (EU) General Product Safety Regulation (GPSR) is a set of rules that requires consumer products to be safe and our obligations to ensure this.

If you have any concerns about our products, you can contact us on ProductSafety@springernature.com

In case Publisher is established outside the EU, the EU authorized representative is:

Springer Nature Customer Service Center GmbH
Europaplatz 3
69115 Heidelberg, Germany

FSC
www.fsc.org
MIX
Papier | Fördert gute Waldnutzung
FSC® C083411

Kolibri 360 GmbH
Ferdinand-Jühlke-Straße 7
99095 Erfurt, Deutschland
produktsicherheit@kolibri360.de